Helicobacter Infection in Clinical Practice

Byung-Wook Kim
Editor

Helicobacter Infection in Clinical Practice

Springer

Editor
Byung-Wook Kim
Department of Internal Medicine
Incheon St. Mary's Hospital
Seoul, Korea (Republic of)

ISBN 978-981-97-8399-1 ISBN 978-981-97-8397-7 (eBook)
https://doi.org/10.1007/978-981-97-8397-7

This Springer imprint is published by the registered company Springer Nature Singapore Pte Ltd.
The registered company address is: 152 Beach Road, #21-01/04 Gateway East, Singapore 189721, Singapore

Preface

Helicobacter pylori (*H. pylori*) is linked to the development of various gastroduodenal diseases, including chronic gastritis, peptic ulcer disease, gastric adenocarcinoma, and gastric mucosa-associated lymphoid tissue (MALT) lymphoma. The interaction between the bacterium and the host immune response may contribute to these different outcomes. In Korea, my homeland, gastric adenocarcinoma is the fourth most common cancer, with about 30,000 new cases registered in National Cancer Registration Program every year. Fortunately, the annual incidence of gastric adenocarcinoma has been declining over the past few decades in Korea due to increased *H. pylori* eradication and the endoscopic resection of gastric adenomas.

Many gastroenterologists and infectious diseases specialists have struggled to enhance the eradication rate of this bacterium. Despite these efforts, the eradication rate of *H. pylori* still does not meet the Maastricht/Florence guidelines in various areas, including Korea. The development of new antibiotics for clinical use has been too slow to keep pace with clinical needs. Strong acid suppressants, like potassium competitive acid blockers instead of proton pump inhibitors, or adding bismuth to the eradication regimen, might offer new hope in the future. Switching from empirical therapy to tailored therapy can also markedly increase the first-line eradication rate, and tailored therapy is cost-effective although it requires more time.

As a clinician, I have found that there are not enough clinically helpful books available in the real world. Therefore, I decided to write this book after a suggestion from Springer. Highlights in each chapter summarize the key points, which I believe will be helpful for readers who do not have enough time to read all the contents of the book. Also, most clinicians are interested in the diagnosis and treatment of *H. pylori*, and this book focuses on these aspects.

My colleagues, who are specialists in *H. pylori* research, participated in writing this book, including Professor Sun Young Lee and Dr. Chul Min Shin. Professor Sun Young Lee provided valuable endoscopy pictures for this book. Dr. Joon Sung Kim endeavored to recruit and edit the manuscripts, and I greatly appreciate his efforts. I would like to express my sincere gratitude to all of the authors and illustrators.

Seoul, Korea (Republic of) Byung-Wook Kim

Contents

Part III Treatment of *Helicobacter pylori*

Part IV Guidelines for *Helicobacter pylori*

Part I

Introduction

1 When to Diagnose and Treat *Helicobacter pylori*?

Byung-Wook Kim

Highlights

What is known
• When a person is suspected of being infected with *H. pylori*, appropriate diagnostic tests should be applied.
• Performing a diagnostic test for *H. pylori* without any plans for eradicating the bacteria is considered unethical.
• The eradication of *H. pylori* is recommended even in the absence of symptoms.
Future research questions
• Is *H. pylori* eradication adequate, even in the presence of advanced atrophic gastritis and intestinal metaplasia?
• Is *H. pylori* eradication cost-effective in regions with a low incidence of gastric cancer?

1.1 Introduction

It has been reported that approximately half of the global population is affected by *H. pylori* with regional variations [1]. Despite the high infection rate reported in numerous studies worldwide, a significant proportion of infected individuals will remain asymptomatic. Nevertheless, it is widely recognized that *Helicobacter pylori* is linked to the development of various gastroduodenal diseases, including chronic gastritis, peptic ulcer disease, gastric adenocarcinoma, and gastric mucosa-associated lymphoid tissue (MALT) lymphoma [2–4]. This chapter provides an overview of the fundamental mechanisms of *H. pylori* infection, the diseases associated with *H. pylori*, and the optimal timing for the diagnosis and treatment of *H. pylori*.

B.-W. Kim (✉)
Department of Internal Medicine, Incheon St. Mary's Hospital, The Catholic University of Korea, Seoul, Korea
e-mail: gastro@catholic.ac.kr

1.2 Main Body

1.2.1 Route of Transmission

Humans serve as the primary reservoir of *Helicobacter pylori;* however, the exact mechanism of transmission between individuals remains uncertain. Several studies have provided support for person-to-person transmission of *H. pylori* through various routes, including gastro-oral, fecal-oral, or oral-oral exposure [5, 6]. One study identified sibling-to-sibling transmission as evidenced by the increased risk of infection associated with household crowding and living with three or more children [7]. Additionally, intimate contact between mother and child has also been suggested as a potential transmission route [8]. It is worth noting that natural transmission could occur through contact with infected vomitus or by consuming *H. pylori*-contaminated water [9, 10].

B.-W. Kim (ed.), *Helicobacter Infection in Clinical Practice*,
https://doi.org/10.1007/978-981-97-8397-7_1

1.2.2 Colonization of *H. pylori*

H. pylori primarily colonizes gastric mucosa. However, the bacteria can also colonize other gastric-type mucosa, such as the ectopic gastric mucosa of the esophagus, gastric metaplasia of the duodenum, and gastric-type Meckel's diverticulum [11–13]. Upon colonization of the gastric mucosa, *H. pylori* utilizes its motility, phospholipase activity, and mucinase activity to penetrate the mucous layer and eventually anchor itself to the epithelial layer [14]. Nonetheless, it is important to note that a proportion of the *H. pylori* exists in a non-adherent state.

1.2.3 Interaction Between Host and *H. pylori*

Gastric epithelial cells play an integral part in the host's response to *H. pylori* infection by encountering various challenges, including antimicrobial factors present in the mucous secretions of the stomach. To survive in the hostile environment, *H. pylori* must evade the immune response. The bacteria can accomplish this by circumventing the antimicrobial effects of mucous secretion factors like lactoferrin, lysozyme, and defensins [15–17].

The host's immune or inflammatory response to *H. pylori* infection significantly contributes to the development of gastrointestinal diseases. In particular, the activation of Th1 cells induced by *H. pylori* infection can trigger apoptosis in gastric epithelial cells [18]. Moreover, Th1 cells can promote the recruitment and activation of neutrophils, B cells, mast cells, and macrophages, potentially leading to structural and functional damage to the stomach [19, 20]. Some reports have suggested that the Th1 or Th17 subsets may be associated with the development of gastrointestinal diseases such as peptic ulcer disease or gastric cancer [21–23]. However, the pathogenesis of gastrointestinal diseases resulting from *H. pylori* infection is intricate, involving various inflammatory cytokines and immune mechanisms.

Despite certain limitations, several epidemiologic studies have demonstrated a protective association between *H. pylori* infection and the incidence of allergic diseases, including asthma, allergic rhinitis, and allergic dermatitis [24–26]. The underlying mechanism of this protective effect remains unclear requires further investigation. Chapter 2 will provide additional insights into *H. pylori* and its associated diseases.

1.2.4 Optimal Timing of Diagnosis

Various diagnostic markers have been developed to identify *H. pylori* infection, and the optimal timing for each method can vary. Direct markers involve the identification of the bacteria using special stains like Warthin–Starry stains and modified Giemsa stains or through culture, providing evidence of the presence of *H. pylori*. The urea breath test relies on the organism's urease activity, and a positive test indicates a current *H. pylori* infection. On the other hand, indirect markers such as serum IgG antibodies to *H. pylori* require a considerable amount of time after infection before they become detectable. Additionally, serologic markers such as serum gastrin and pepsinogen I and II indicate the extent of gastric mucosal atrophy and cannot be used as evidence of acute *H. pylori* infection.

When a person is suspected of being infected with *H. pylori*, appropriate diagnostic tests should be applied. Performing a diagnostic test for *H. pylori* without any plans for eradicating the bacteria is considered unethical. Further details on diagnostic methods for *H. pylori* will be discussed in Chap. 3.

1.2.5 Optimal Timing of Treatment

The Kyoto consensus introduced an etiology-based classification for gastritis, leading to the inclusion of *H. pylori* gastritis as a distinct disease entity in ICD 11 [27]. Consequently, the eradication of *H. pylori* is recommended even in the absence of symptoms, according to various guidelines [27–30]. Most guidelines strongly advocate for *H. pylori* eradication in cases of peptic ulcer disease, after endoscopic resection for gastric adenocarcinoma, and in MALT lymphoma [28–31]. *H. pylori* eradication treatment reduced the risk of gastric cancer in first-degree relatives of gastric cancer [32].

Interestingly, there have been reports suggesting that *H. pylori* infection may reduce the risk of asthma and allergic diseases [33]. Additionally, some studies have shown that *H. pylori* can significantly decrease the risk of esophageal adenocarcinoma in Western populations [34]. Despite these findings, it is important to acknowledge the limitations of these studies. A summary of indications from various guidelines can be found in Table 1.1.

However, consensus regarding the optimal timing of treatment is lacking. When a person has diseases with absolute indications for *H. pylori* eradication therapy, immediate treatment is recommended. Nevertheless, debates surround the effectiveness of *H. pylori* eradication therapy in preventing gastric cancer. A recent meta-analysis indicated a 46% reduction in the risk of gastric cancer following *H. pylori* eradication, with a number needed to treat of 72 individuals to prevent one case of gastric cancer [35]. To prevent gastric cancer, *H. pylori* eradication should be prioritized among younger adults residing in high

Table 1.1 Indications of *H. pylori* eradication from various guidelines

		Maastricht VI/ Florence [28]	ACG [29]	Japan [30]	Korea [31]
Classic indications	Peptic ulcer disease	Definite	Definite	Definite	Definite
	Endoscopic resection for early gastric cancer	Definite	Definite	Definite	Definite
	Gastric MALT lymphoma	Definite	Definite	Definite	Definite
Expanded indications	ITP	Definite	Moderate	Definite	Definite
	Unexplained IDA	Definite	Moderate	Definite	Moderate
	Family history of gastric cancer	Subtle	N/A	N/A	Definite
	Long-term low-dose aspirin user	Moderate	Moderate	N/A	Definite
	Long-term NSAID user	Moderate	Definite	N/A	N/A
	Endoscopic resection for gastric adenoma	N/A	N/A	N/A	Moderate
Gastric cancer prevention	General population	Moderate	N/A	Definite	N/A
	Atrophic gastritis	Definite	N/A	Definite	Subtle
	Intestinal metaplasia	Definite	N/A	Definite	Subtle
Other upper gastrointestinal diseases	Functional dyspepsia	Subtle	Definite	Definite	Definite
	GERD	N/A	No	Definite	N/A
	Hyperplastic gastric polyp	N/A	No	Definite	N/A
	Diffuse large B-cell lymphoma of stomach	N/A	N/A	Unknown	N/A
Lower gastrointestinal diseases	Rectal MALT lymphoma	N/A	N/A	No	N/A
	Cap polyposis	N/A	N/A	Unknown	N/A
Other diseases	DM	N/A	N/A	Moderate	N/A
	Vitamin B12 deficiency	Definite	N/A	N/A	N/A
	Chronic urticaria	N/A	N/A	Moderate	N/A
	Parkinson's syndrome	N/A	N/A	Unknown	N/A
	Alzheimer's disease	N/A	N/A	Unknown	N/A

MALT mucosa-associated lymphoid tissue, *ITP* idiopathic thrombocytopenic purpura, *IDA* iron-deficiency anemia, *NSAID* nonsteroidal anti-inflammatory drug, *GERD* gastroesophageal reflux disease, *DM* diabetes mellitus, *N/A* not available, *ACG* American College of Gastroenterology incidence regions, before the onset of atrophic gastritis and intestinal metaplasia [36].

1.3 Conclusions

H. pylori can be transmitted through various routes, such as gastro-oral, fecal-oral, or oral-oral exposure. It primarily colonizes the gastric mucosa and triggers immune or inflammatory reactions in the host, contributing to the development of gastrointestinal diseases. Several direct and indirect methods have been developed to detect *H. pylori* infection, and the optimal timing for each method may vary. Most guidelines strongly recommend *H. pylori* eradication in cases of peptic ulcer disease, after endoscopic resection for gastric adenocarcinoma, and in MALT lymphoma. To effectively prevent gastric cancer, it is crucial to prioritize *H. pylori* eradication among younger adults residing in high incidence regions, before the onset of atrophic gastritis and intestinal metaplasia.

References

1. Hooi JKY, Lai WY, Ng WK, et al. Global prevalence of *Helicobacter pylori* infection: Systematic review and meta-analysis. Gastroenterology. 2017;153:420–9.
2. Peek RM Jr, Blaser MJ. *Helicobacter pylori* and gastrointestinal tract adenocarcinomas. Nat Rev Cancer. 2002;2:28–37.
3. Blaser MJ. *Helicobacter pylori* and gastric diseases. BMJ. 1998;316:1507–10.
4. Wroblewski LE, Peek RM Jr, Wilson KT. *Helicobacter pylori* and gastric cancer: factors that modulate disease risk. Clin Microbiol Rev. 2010;23:713–39.
5. Brown LM. *Helicobacter pylori*: epidemiology and routes of transmission. Epidemiol Rev. 2000;22:283–97.
6. Amieva MR, El-Omar EM. Host-bacterial interactions in *Helicobacter pylori* infection. Gastroenterology. 2008;134:306–23.
7. Porras C, Nodora J, Sexton R, et al. Epidemiology of *Helicobacter pylori* infection in six Latin American countries (SWOG Trial S0701). Cancer Causes Control. 2013;24:209–15.
8. Weyermann M, Adler G, Brenner H, et al. The mother as source of *Helicobacter pylori* infection. Epidemiology. 2006;17:332–4.
9. Perry S, de la Luz SM, Yang S, et al. Gastroenteritis and transmission of *Helicobacter pylori* infection in households. Emerg Infect Dis. 2006;12:1701–8.
10. Bellack NR, Koehoorn MW, MacNab YC, et al. A conceptual model of water's role as a reservoir in *Helicobacter pylori* transmission: a review of the evidence. Epidemiol Infect. 2006;134:439–49.
11. Martins FP, Artigiani Neto R, Oshima CT, et al. Overexpression of cyclooxygenase-2 in endoscopic biopsies of ectopic gastric mucosa. Braz J Med Biol Res. 2007;40:1447–54.
12. Shabib SM, Cutz E, Drumm B, et al. Association of gastric metaplasia and duodenitis with *Helicobacter pylori* infection in children. Am J Clin Pathol. 1994;102:188–91.
13. Bemelman WA, Bosma A, Wiersma PH, et al. Role of *Helicobacter pylori* in the pathogenesis of complications of Meckel's diverticula. Eur J Surg. 1993;159:171–5.
14. Testerman TL, McGee DJ, Mobley HLT. Adherence and colonization. In: Mobley HLT, Mendz GL, Hazell SL, editors. *Helicobacter pylori*: physiology and genetics. Washington DC: ASM Press; 2001. p. 379–417.
15. Nakao K, Imoto I, Gabazza EC, et al. Gastric juice levels of lactoferrin and *Helicobacter pylori* infection. Scand J Gastroenterol. 1997;32:530–4.
16. Negrini R, Lisato L, Zanella I, et al. *Helicobacter pylori* infection induces antibodies cross-reacting with human gastric mucosa. Gastroenterology. 1991;101:437–45.
17. O'Neil DA, Cole SP, Martin-Porter E, et al. Regulation of human beta-defensins by gastric epithelial cells in response to infection with *Helicobacter pylori* or stimulation with interleukin-1. Infect Immun. 2000;68:5412–5.
18. Wagner S, Beil W, Westermann J, et al. Regulation of epithelial cell growth by *Helicobacter pylori*: evidence for a major role of apoptosis. Gastroenterology. 1997;113:1836–47.
19. Crabtree JE. Role of cytokines in pathogenesis of *Helicobacter pylori*-induced mucosal damage. Dig Dis Sci. 1998;43:46S–55S.
20. Moss SF. The carcinogenic effect of *H. pylori* on the gastric epithelial cell. J Physiol Pharmacol. 1999;50:847–56.
21. Robinson K, Kenefeck R, Pidgeon EL, et al. *Helicobacter pylori*-induced peptic ulcer disease is associated with inadequate regulatory T cell responses. Gut. 2008;57:1375–85.
22. Stromberg E, Lundgren A, Edebo A, et al. Increased frequency of activated T-cells in the *Helicobacter pylori*-infected antrum and duodenum. FEMS Immunol Med Microbiol. 2003;36:159–68.
23. Caruso R, Fina D, Paoluzi OA, et al. IL-23-mediated regulation of IL-17 production in *Helicobacter pylori*-infected gastric mucosa. Eur J Immunol. 2008;38:470–8.

24. Blaser MJ, Chen Y, Reibman J. Does *Helicobacter pylori* protect against asthma and allergy? Gut. 2008;57:561–7.
25. Kosunen TU, Hook-Nikanne J, Salomaa A, et al. Increase of allergen-specific immunoglobulin E antibodies from 1973 to 1994 in a Finnish population and a possible relationship to *Helicobacter pylori* infections. Clin Exp Allergy. 2002;32:373–8.
26. Pessi T, Virta M, Adjers K, et al. Genetic and environmental factors in the immunopathogenesis of atopy: interaction of *Helicobacter pylori* infection and IL4 genetics. Int Arch Allergy Immunol. 2005;137:282–8.
27. Sugano K, Tack J, Kuipers EJ, et al. Kyoto global consensus report on *Helicobacter pylori* gastritis. Gut. 2015;64:1353–67.
28. Malfertheiner P, Megraud F, Rokkas T, et al. Management of *Helicobacter pylori* infection: the Maastricht VI/Florence consensus report. Gut 2022:gutjnl-2022-327745. https://doi.org/10.1136/gutjnl-2022-327745. Online ahead of print.
29. Chey WD, Leontiadis GI, Howden CW, et al. ACG clinical guideline: treatment of *Helicobacter pylori* infection. Am J Gastroenterol. 2017;112:212–39.
30. Kato M, Ota H, Okuda M, et al. Guidelines for the management of *Helicobacter pylori* infection in Japan: 2016 Revised Edition. Helicobacter. 2019;24:e12597.
31. Jung HK, Kang SJ, Lee YC, et al. Evidence based guidelines for the treatment of *Helicobacter pylori* infection in Korea 2020. Korean J Intern Med. 2021;36:807–38.
32. Choi IJ, Kim CG, Lee JY, et al. Family history of gastric cancer and *Helicobacter pylori* treatment. N Engl J Med. 2020;382:427–36.
33. Ness-Jensen E, Langhammer A, Hveem K, et al. *Helicobacter pylori* in relation to asthma and allergy modified by abdominal obesity: the HUNT study in Norway. World Allergy Organ J. 2019;12:100035.
34. Xie FJ, Zhang YP, Zheng QQ, et al. *Helicobacter pylori* infection and esophageal cancer risk: an updated meta-analysis. World J Gastroenterol. 2013;19:6098–107.
35. Ford AC, Yuan Y, Moayyedi P. *Helicobacter pylori* eradication therapy to prevent gastric cancer: systematic review and meta-analysis. Gut. 2020;69:2113–21.
36. Liou JM, Malfertheiner P, Lee YC, et al. Screening and eradication of *Helicobacter pylori* for gastric cancer prevention: the Taipei global consensus. Gut. 2020;69:2093–112.

Helicobacter pylori-Related Diseases

2

Byung-Wook Kim

Highlights

What is known

- *H. pylori* is responsible for causing various gastric diseases such as chronic atrophic gastritis, peptic ulcer disease, gastric adenocarcinoma, and gastric MALToma.
- *H. pylori* is responsible for causing various extra-gastric diseases.

Future research questions

- Evidence regarding the relationship between *H. pylori* infection and each disease entity should be clarified.
- The advantages of *H. pylori* eradication in these diseases should be clarified.

2.1 Introduction

When *Helicobacter pylori* colonizes the gastric mucosa, it continuously damages the mucosa, leading to various diseases, including acute and chronic gastritis, peptic ulcer diseases, gastric adenocarcinoma, and gastric marginal zone B-cell lymphoma of MALT type (MALT lymphoma) [1, 2]. The diverse pathologic outcomes primarily arise from a combination of interactions between the host, bacterial virulence, and environmental factors [3, 4]. Additionally, *H. pylori* is responsible for causing extra-gastric diseases, including idiopathic thrombocytopenic purpura (ITP). This chapter will focus on discussing *H. pylori*-related diseases.

2.2 Main Body

2.2.1 Gastritis

H. pylori colonizes gastric mucosa, especially in the antrum, and advances to the corpus. This concept is exactly matching with Kimura-Takemoto's classification of atrophic gastritis [5]. Operative link for gastritis assessment (OLGA) staging system for gastric cancer risk stratification is also based on this concept [6]. The molecular mechanisms how *H. pylori* locate and colonize the gastric epithelium have not been clearly elucidated so far. One study suggested that *H. pylori* is attracted by urea emanated from gastric type epithelial cells, which is detected by the chemoreceptor TlpB [7].

About 80% of *H. pylori*-infected persons remain asymptomatic. Colonization of *H. pylori* at the gastric mucosa is thought to be introduced during the early years of life and persist indefinitely unless treated [8]. Correa suggested a cascade model on gastritis to gastric adenocarcinoma (Fig. 2.1). *H. pylori* plays a key role in each step.

B.-W. Kim (✉)
Department of Internal Medicine, Incheon St. Mary's Hospital, The Catholic University of Korea, Seoul, Korea
e-mail: gastro@catholic.ac.kr

B.-W. Kim (ed.), *Helicobacter Infection in Clinical Practice*,
https://doi.org/10.1007/978-981-97-8397-7_2

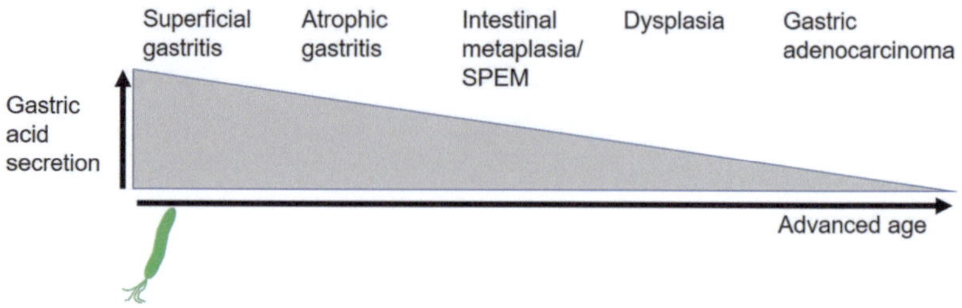

Fig. 2.1 Correa cascade model on gastritis to gastric adenocarcinoma

2.2.2 Peptic Ulcer Diseases

Ulcers occur within the stomach and/or duodenum and are often chronic in nature. The global prevalence of peptic ulcer disease (PUD) is declining in parallel with a decreasing prevalence of *H. pylori* infection [9]. However, nonsteroidal anti-inflammatory drug (NSAID)-related PUD is increasing. Despite changing global trends in PUD, *H. pylori* remains the most prevalent cause of PUD.

H. pylori has the potential to damage the mucosal defense system through the reduction in thickness of the mucus gel layer or the decrease in mucosal blood flow [10]. Additionally, it can stimulate an increase in gastric acid secretion by producing diverse antigens, virulence factors, and soluble mediators. These factors collectively lead to an augmentation in parietal-cell mass. The interplay between the host and *H. pylori*, along with the mucosal inflammatory response, may have a genetic basis and play a role in determining the outcomes of conditions like PUD and other acid-related diseases [11].

Functional polymorphisms in various cytokine genes have been associated with the occurrence of PUD. Interleukin 1β (IL-1β), which is encoded by *IL-1B*, represents a cytokine linked to both the inflammatory response to *H. pylori* infection and the inhibition of gastric secretion [11–13]. Furthermore, genes encoding tumor necrosis factor (TNF) and lymphotoxin-α have demonstrated association with duodenal ulcers and gastric ulcers [14, 15].

The inflammation associated with *H. pylori* infection can result in either hypochlorhydria or hyperchlorhydria, ultimately determining the specific type of PUD that forms [16]. However, a comprehensive understanding of how *H. pylori* infection leads to the development of different types of PUD is not yet fully attained.

2.2.3 Gastric Adenocarcinoma

It is well known that intestinal type gastric adenocarcinoma by Lauren's classification is more closely related to *H. pylori* infection compared to diffuse type. Development of gastric cancer was proven by animal model studies, especially in Mongolian gerbils [17, 18]. Subjects infected by *H. pylori* with *cagA* positive strain have a higher risk of developing a PUD or gastric adenocarcinoma than those infected with *cagA* negative strain [19].

The dysbiotic gastric microbiome is an emerging concept linking chronic inflammation with gastric neoplasia [20–22]. Infection with *H. pylori* can lead to alterations in the normal gastric microbiome. Comparing individuals with and without *H. pylori* infection, a reduction in gastric microbial diversity is evident, as seen in the previous studies [23, 24]. Notably, a study revealed elevated levels of *Proteobacteria, Firmicutes, Fusobacteria, Actinobacteria*, and *Bacteroidetes* in the gastric mucosa of patients with gastric cancer [25].

2.2.4 Gastric Marginal Zone B-Cell Lymphoma of Mucosa-Associated Lymphoid Tissue Lymphoma (MALT), Gastric MALT Lymphoma

Marginal zone lymphoma arises from B-cells originating after the peripheral zone of the germinal center known as the marginal zone. Gastric

MALT lymphoma is associated with *H. pylori* infection in 85~90% of cases [26]. This infection triggers a Th1-type immune response, characterized by the involvement of multiple pro-inflammatory cytokines. The activation and proliferation of B-cells are strongly influenced by signals from *H. pylori*-specific CD4+ T helper cells, primarily through CD40-mediated signaling and Th2 cytokines [27]. Additionally, cytotoxic T lymphocytes within the marginal zone exhibit impaired cytotoxicity activity involving perforin and a diminished capability to induce apoptosis through the Fas-Fas ligand system. These mechanisms collectively support B-cell expansion and contribute to the development of lymphoma [28].

2.2.5 Extra-Gastric Diseases (Fig. 2.2)

2.2.5.1 Gastroesophageal Reflux Disease

Population-based observational studies have reported a negative correlation between *H. pylori* infection and gastroesophageal reflux disease (GERD) [29–31]. The pathophysiologic mechanism is yet to be fully understood, but it is plausible that *H. pylori* infection induces atrophy in the corpus and antrum of the stomach, leading to a reduction in gastric acid secretion [32]. The clinical significance of this finding lies in determining whether the eradication of *H. pylori* can worsen GERD. Fortunately, there is no definite evidence suggesting that the eradication of *H. pylori* exacerbates GERD [33–35].

2.2.5.2 Colorectal Cancer

According to a cohort study conducted in Germany [36] and a study of African-Americans in the United States [37], the risk of colorectal cancer in the *H. pylori*-infected group was 1.85 (95% CI 1.14–2.99) and 1.5 (95% CI 1.2–1.9), respectively. However, a meta-analysis of nine East Asian studies found no significant association between *H. pylori* infection and colorectal cancer [38]. Some researchers proposed that increased serum gastrin levels and cyclooxygenase expression following *H. pylori* infection might impact the development of colorectal cancer [39], although further studies are needed to clarify the underlying pathophysiologic mechanism.

2.2.5.3 Hematologic Disorders

1. Idiopathic thrombocytopenic purpura

Among the associations between *H. pylori* infection and non-gastrointestinal tract diseases, it is the most actively researched area. Gasbarrini et al. reported that among 11 patients with autoimmune thrombocytopenia, 8 patients showed improved platelet counts after *H. pylori* eradication, and 6 patients no longer had autoantibodies against platelets [40]. Since then, numerous studies have been conducted to explore this phenomenon. The underlying pathophysiological mechanism involves cross-molecular similarity between the *cagA* protein of *H. pylori* and platelet antigen in certain patients with idiopathic thrombocytopenic purpura (ITP), leading to an immune response. Consequently, this immune reaction may result in a decrease in platelet count [41]. A meta-analysis revealed that approximately half of the patients with ITP demonstrated an improvement in platelet count following *H. pylori* eradication [42].

2. Iron-deficiency anemia

The mechanisms by which *H. pylori* infection contributes to iron-deficiency anemia (IDA) involve multiple factors. These include blood loss through erosive lesions of the gastric mucosa caused by *H. pylori* infection, reduced iron absorption due to decreased gastric acid secretion, iron consumption by *H. pylori*, and the possibility of anemia resulting from chronic inflammation [43–46]. Several studies, including researches on Korean adolescents and population-based studies in the West, have shown that *H. pylori*-infected individuals have significantly lower serum ferritin levels compared to noninfected individuals [47–49]. A meta-analysis of prospective studies investigat-

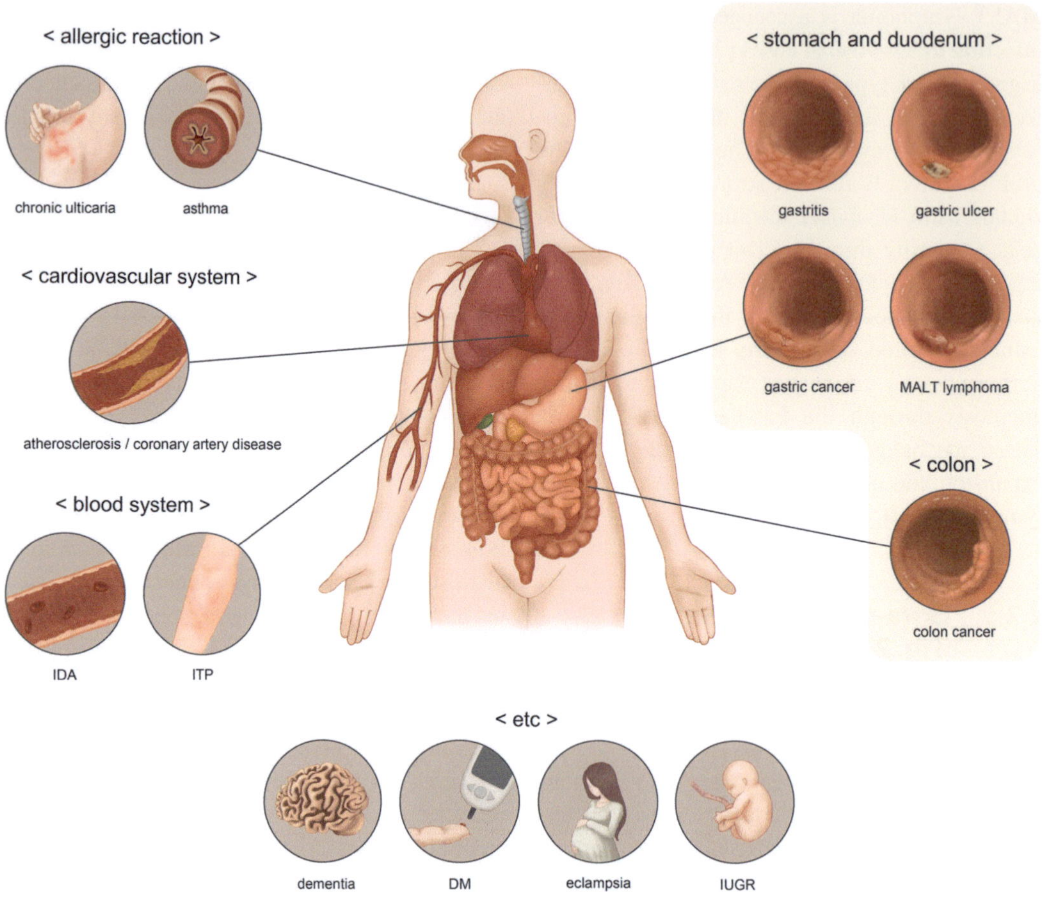

Fig. 2.2 Gastric and extra-gastric diseases related to *H. pylori* infection

ing the impact of *H. pylori* eradication on hemoglobin levels revealed that the group receiving both iron administration and *H. pylori* eradication exhibited significantly elevated serum iron and ferritin levels compared to the group receiving iron administration alone [50].

2.2.5.4 Cardiovascular Disorders

1. Atherosclerosis

Inflammation of the arteries is a key factor in the development of atherosclerosis, and certain infections, including *Chlamydia pneumoniae*, cytomegalovirus, and *H. pylori*, have been associated with this condition [51]. The mechanism by which *H. pylori* infection contributes to the development of atherosclerosis involves sustaining a low systemic inflammatory response, which in turn promotes the production of cytokines such as C-reactive protein (CRP), interleukin (IL)-6, and IL-18 [52–54]. These cytokines play a significant role in the induction of atherosclerosis.

2. Coronary heart disease

A meta-analysis demonstrated that chronic *H. pylori* infection increases the risk of early development of coronary heart disease (CHD), especially in the general population [55]. Other studies support this fact, as anti-*H. pylori* immunoglobulin G (IgG) has been associated with an increased risk of CHD [56]. Cytotoxin-associated gene-A (Cag-A) antibodies might play a role in the development of CHD, considering that patients with *H. pylori* infection have a higher prevalence of positive Cag-A antibodies [57, 58]. However, clear underlying mechanism has not been elucidated so far.

2.2.5.5 Allergic Diseases

1. Asthma

It is recognized that there is an inverse correlation between *H. pylori* infection and the development of bronchial asthma. Two meta-analyses showed a weak inverse correlation between *H. pylori* infection and asthma [59, 60]. In an animal experiment, when newborn mice were infected with the *H. pylori*, the immune response to control the infection did not occur properly, and as a result, the immune response was not regulated, suppressing the development of diseases related to the immune response such as asthma [61]. That is, when Th1 is induced by *H. pylori* infection, an immunologic deviation occurs and suppresses the Th2 response induced by an allergen [62].

2. Chronic urticaria

It has been reported that symptoms improved after eradication of *H. pylori* in patients with chronic urticaria, but the pathophysiological mechanism is still unclear, so basic research is needed [63]. There was also a study on the association between *H. pylori* infection and rosacea [64].

2.2.5.6 Others

1. Dementia

Studies have reported a potential link between *H. pylori* infection and the development of vascular dementia [65, 66]. The infection is believed to contribute to this condition by promoting the production of inflammatory cytokines, including IL-1β, IL-6, and tumor necrosis factor-α (TNF-α), which can lead to continuous damage to cerebral blood vessel [66]. Additionally, there have been suggestions of an association between *H. pylori* infection and tauopathies like Alzheimer's disease, possibly mediated by urease produced by *H. pylori* [67].

2. Diabetes

H. pylori infection has been identified as a potential risk factor for diabetes. Specifically, studies have shown that *H. pylori* infection can increase hemoglobin A1c levels in Asian patients aged 65 or older with type 2 diabetes [68]. This may be attributed to the induction of insulin resistance through an inflammatory response following *H. pylori* infection. Furthermore, a meta-analysis revealed that *H. pylori* infection was associated with an increased risk of diabetic complications, including nephropathy (relative risk 1.35, 95% CI 1.06–1.73) and neuropathy (relative risk 1.73, 95% CI 1.03–1.40) [69].

3. Impact on pregnancy

There was a report that *H. pylori* infection increased the risk of pre-eclampsia, and the relative risk was 2.67 (95% CI 1.08–6.57) [70]. Although the pathophysiological mechanism is not precise, it has been suggested that *H. pylori* infection induces an inflammatory response and oxidative stress, resulting in endothelial dysfunction resulting in pre-eclampsia [71].

Fetal growth retardation has been found to be more prevalent among mothers infected with *H. pylori* [72]. The underlying pathophysiological mechanism involves a higher incidence of peptic ulcer disease and dyspepsia in *H. pylori*-infected mothers, which can lead to reduced food intake during pregnancy [73]. Furthermore, anemia is more frequently observed in *H. pylori*-infected mothers, and this can contribute to fetal growth retardation [71].

2.3 Conclusions

The involvement of *H. pylori* in the development of gastroduodenal diseases is indisputable. Other organs such as hematologic, cardiovascular, and central nervous systems also might be affected by

H. pylori infection. However, the advantages of *H. pylori* eradication in these diseases should be clarified through further studies.

References

1. Suerbaum S, Michetti P. *Helicobacter pylori* infection. N Engl J Med. 2002;347:1175–86.
2. Peek RM, Blaser MJ. *Helicobacter pylori* and gastrointestinal tract adenocarcinomas. Nat Rev Cancer. 2002;2:28–37.
3. Amieva MR, El-Omar EM. Host-bacterial interactions in *Helicobacter pylori* infection. Gastroenterology. 2008;134:306–23.
4. van Amsterdam K, van Vliet AH, Kusters JG, et al. Of microbe and man: determinants of *Helicobacter pylori*-related diseases. FEMS Microbiol Rev. 2006;30:131–56.
5. Kimura K, Takemoto T. An endoscopic recognition of the atrophic border and its significance in chronic gastritis. Endoscopy. 1969;1:87–97.
6. Rugge M, Meggio A, Pennelli G, et al. Gastritis staging in clinical practice: the OLGA staging system. Gut. 2007;56:631–6.
7. Huang JY, Sweeney EG, Sigal M, et al. Chemodetection and destruction of host urea allows *Helicobacter pylori* to locate the epithelium. Cell Host Microbe. 2015;18:147–56.
8. Plummer M, Franceschi S, Vignat J, et al. Global burden of gastric cancer attributable to *Helicobacter pylori*. Int J Cancer. 2015;136:487–90.
9. Hooi JKY, Lai WY, Ng WK, et al. Global prevalence of *Helicobacter pylori* infection: systematic review and meta-analysis. Gastroenterology. 2017;153:420–9.
10. Yuan Y, Padol IT, Hunt RH. Peptic ulcer disease today. Nat Clin Pract Gastroenterol Hepatol. 2006;3:80–9.
11. Datta D, Roychoidhury S. To be or not to be: the host genetic factor and beyond *Helicobacter pylori* mediated gastro-duodenal diseases. World J Gastroenterol. 2015;21:2883–95.
12. Furuta T, El-Omar EM, Xiao F, et al. Interleukin 1β polymorphisms increase risk of hypochlorhydria and atrophic gastritis and reduce risk of duodenal ulcer recurrence in Japan. Gastroenterology. 2002;123:92–105.
13. Chakravorty M, Ghosh A, Choudhury A, et al. Interaction between IL1B gene promoter polymorphisms in determining susceptibility to *Helicobacter pylori* associated duodenal ulcer. Hum Mutat. 2006;27:411–9.
14. Zhang BB, Liu XZ, Sun J, et al. Association between TNF α gene polymorphisms and the risk of duodenal ulcer: a meta-analysis. PLoS One. 2013;8:e57167.
15. Lanas A, García-González MA, Santolaria S, et al. TNF and LTA gene polymorphisms reveal different risk in gastric and duodenal ulcer patients. Genes Immun. 2001;2:415–21.
16. Meyer-Rosberg K, Scott DR, Rex D, et al. The effect of environmental pH on the proton motive force of *Helicobacter pylori*. Gastroenterology. 1996;111:886–900.
17. Watanabe T, Tada M, Nagai H, et al. *Helicobacter pylori* infection induces gastric cancer in mongolian gerbils. Gastroenterology. 1998;115:642–8.
18. Honda S, Fujioka T, Tokieda M, et al. Development of *Helicobacter pylori*-induced gastric carcinoma in Mongolian gerbils. Cancer Res. 1998;58:4255–9.
19. Yamaoka Y. Mechanisms of disease: *Helicobacter pylori* virulence factors. Nat Rev Gastroenterol Hepatol. 2010;7:629–41.
20. Ferreira RM, Pereira-Marques J, Pinto-Ribeiro I, et al. Gastric microbial community profiling reveals a dysbiotic cancer-associated microbiota. Gut. 2018;67:226–36.
21. Guo Y, Zhang Y, Gerhard M, et al. Effect of *Helicobacter pylori* on gastrointestinal microbiota: a population-based study in Linqu, a high-risk area of gastric cancer. Gut. 2020;69:1598–607.
22. Kwon SK, Park JC, Kim KH, et al. Human gastric microbiota transplantation recapitulates premalignant lesions in germ-free mice. Gut. 2022;71:1266–76.
23. Andersson AF, Lindberg M, Jakobsson H, et al. Comparative analysis of human gut microbiota by barcoded pyrosequencing. PLoS One. 2008;3:e2836.
24. Das A, Pereira V, Saxena S, et al. Gastric microbiome of Indian patients with *Helicobacter pylori* infection, and their interaction networks. Sci Rep. 2017;7:1–9.
25. Wang L, Zhou J, Xin Y, et al. Bacterial overgrowth and diversification of microbiota in gastric cancer. Eur J Gastroenterol Hepatol. 2016;28:261–6.
26. Schreuder MI, van den Brand M, Hebeda KM, et al. Novel developments in the pathogenesis and diagnosis of extranodal marginal zone lymphoma. J Hematop. 2017;10:91–107.
27. Hussell T, Isaacson PG, Crabtree JE, et al. The response of cells from low-grade B-cell gastric lymphomas of mucosa-associated lymphoid tissue to *Helicobacter pylori*. Lancet. 1993;342:571–4.
28. D'Elios MM, Amedei A, Manghetti M, et al. Impaired T-cell regulation of B-cell growth in *Helicobacter pylori*-related gastric low-grade MALT lymphoma. Gastroenterology. 1999;117:1105–12.
29. Wu JC, Sung JJ, Chan FK, et al. *Helicobacter pylori* infection is associated with milder gastro-oesophageal reflux disease. Aliment Pharmacol Ther. 2000;14:427–32.
30. Koike T, Ohara S, Sekine H, et al. *Helicobacter pylori* infection prevents erosive reflux oesophagitis by decreasing gastric acid secretion. Gut. 2001;49:330–4.
31. O'Connor HJ. Review article: *Helicobacter pylori* and gastro-oesophageal reflux disease-clinical implications and management. Aliment Pharmacol Ther. 1999;13:117–27.

32. Feldman M, Cryer B, Sammer D, et al. Influence of *H. pylori* infection on meal-stimulated gastric acid secretion and gastroesophageal acid reflux. Am J Physiol. 1999;277:G1159–64.
33. Iijima K, Ohara S, Sekine H, et al. Changes in gastric acid secretion assayed by endoscopic gastrin test before and after *Helicobacter pylori* eradication. Gut. 2000;46:20–6.
34. Moayyedi P, Bardhan C, Young L, et al. *Helicobacter pylori* eradication does not exacerbate reflux symptoms in gastroesophageal reflux disease. Gastroenterology. 2001;121:1120–6.
35. Chung SJ, Lim SH, Choi J, et al. *Helicobacter pylori* serology inversely correlated with the risk and severity of reflux esophagitis in *Helicobacter pylori* endemic area: a matched case-control study of 5,616 health check-up Koreans. J Neurogastroenterol Motil. 2011;17:267–73.
36. Selgrad M, Bornschein J, Kandulski A, et al. *Helicobacter pylori* but not gastrin is associated with the development of colonic neoplasms. Int J Cancer. 2014;135:1127–31.
37. Brim H, Zahaf M, Laiyemo AO, et al. Gastric *Helicobacter pylori* infection associates with an increased risk of colorectal polyps in African Americans. BMC Cancer. 2014;14:296.
38. Guo Y, Li HY. Association between *Helicobacter pylori* infection and colorectal neoplasm risk: a meta-analysis based on East Asian population. J Cancer Res Ther. 2014;10(Suppl):263–6.
39. Hartwich A, Konturek SJ, Pierzchalski P, et al. *Helicobacter pylori* infection, gastrin, cyclooxygenase-2, and apoptosis in colorectal cancer. Int J Colorectal Dis. 2001;16:202–10.
40. Gasbarrini A, Franceschi F, Tartaglione R, et al. Regression of autoimmune thrombocytopenia after eradication of *Helicobacter pylori*. Lancet. 1998;352:878.
41. Kurtoglu E, Kayacetin E, Ugur A. *Helicobacter pylori* infection in patients with autoimmune thrombocytopenic purpura. World J Gastroenterol. 2004;10:2113–5.
42. Franchini M, Cruciani M, Mengoli C, et al. Effect of *Helicobacter pylori* eradication on platelet count in idiopathic thrombocytopenic purpura: a systematic review and meta-analysis. J Antimicrob Chemother. 2007;60:237–46.
43. Annibale B, Capurso G, Lahner E, et al. Concomitant alterations in intragastric pH and ascorbic acid concentration in patients with *Helicobacter pylori* gastritis and associated iron deficiency anaemia. Gut. 2003;52:496–501.
44. Yip R, Limburg PJ, Ahlquist DA, et al. Pervasive occult gastrointestinal bleeding in an Alaska native population with prevalent iron deficiency. Role of *Helicobacter pylori* gastritis. JAMA. 1997;277:1135–9.
45. Sobala GM, Crabtree JE, Dixon MF, et al. Acute *Helicobacter pylori* infection: clinical features, local and systemic immune response, gastric mucosal histology, and gastric juice ascorbic acid concentrations. Gut. 1991;32:1415–8.
46. Barabino A, Dufour C, Marino CE, et al. Unexplained refractory iron-deficiency anemia associated with *Helicobacter pylori* gastric infection in children: further clinical evidence. J Pediatr Gastroenterol Nutr. 1999;28:116–9.
47. Choe YH, Kim SK, Hong YC. *Helicobacter pylori* infection with iron deficiency anaemia and subnormal growth at puberty. Arch Dis Child. 2000;82:136–40.
48. Milman N, Rosenstock S, Andersen L, et al. Serum ferritin, hemoglobin, and *Helicobacter pylori* infection: a seroepidemiologic survey comprising 2794 Danish adults. Gastroenterology. 1998;115:268–74.
49. Berg G, Bode G, Blettner M, et al. *Helicobacter pylori* infection and serum ferritin: a population-based study among 1806 adults in Germany. Am J Gastroenterol. 2001;96:1014–8.
50. Yuan W, Yumin L, Kehu Y, et al. Iron deficiency anemia in *Helicobacter pylori* infection: meta-analysis of randomized controlled trials. Scand J Gastroenterol. 2010;45:665–76.
51. Mayr M, Kiechl S, Willeit J, et al. Infections, immunity, and atherosclerosis: associations of antibodies to *Chlamydia pneumoniae*, *Helicobacter pylori*, and cytomegalovirus with immune reactions to heat-shock protein 60 and carotid or femoral atherosclerosis. Circulation. 2000;102:833–9.
52. Kanbay M, Gür G, Yücel M, et al. Does eradication of *Helicobacter pylori* infection help normalize serum lipid and CRP levels? Dig Dis Sci. 2005;50:1228–31.
53. Nakagawa H, Tamura T, Mitsuda Y, et al. Significant association between serum interleukin-6 and *Helicobacter pylori* antibody levels among *H. pylori*-positive Japanese adults. Mediators Inflamm. 2013;2013:142358.
54. Chen BF, Xu X, Deng Y, et al. Relationship between *Helicobacter pylori* infection and serum interleukin-18 in patients with carotid atherosclerosis. Helicobacter. 2013;18:124–8.
55. Sun J, Rangan P, Bhat SS, et al. A meta-analysis of the association between *Helicobacter pylori* infection and risk of coronary heart disease from published prospective studies. Helicobacter. 2016;21:11–23.
56. Rogha M, Nikvarz M, Pourmoghaddas Z, et al. Is *Helicobacter pylori* infection a risk factor for coronary heart disease? ARYA Atheroscler. 2012;8:5–8.
57. Khodaii Z, Vakili H, Ghaderian SM, et al. Association of *Helicobacter pylori* infection with acute myocardial infarction. Coron Artery Dis. 2011;22:6–11.
58. Rozankovic PB, Huzjan AL, Cupic H, et al. Influence of CagA-positive *Helicobacter pylori* strains on atherosclerotic carotid disease. J Neurol. 2011;258:753–61.
59. Wang Q, Yu C, Sun Y. The association between asthma and *Helicobacter pylori*: a meta-analysis. Helicobacter. 2013;18:41–53.

60. Zhou X, Wu J, Zhang G. Association between *Helicobacter pylori* and asthma: a meta-analysis. Eur J Gastroenterol Hepatol. 2013;25:460–8.
61. Arnold IC, Lee JY, Amieva MR, et al. Tolerance rather than immunity protects from *Helicobacter pylori*-induced gastric preneoplasia. Gastroenterology. 2011;140:199–209.
62. Amedei A, Cappon A, Codolo G, et al. The neutrophil-activating protein of *Helicobacter pylori* promotes Th1 immune responses. J Clin Invest. 2006;116:1092–101.
63. Schnyder B, Helbling A, Pichler WJ. Chronic idiopathic urticaria: natural course and association with *Helicobacter pylori* infection. Int Arch Allergy Immunol. 1999;119:60–3.
64. Utaş S, Ozbakir O, Turasan A, et al. *Helicobacter pylori* eradication treatment reduces the severity of rosacea. J Am Acad Dermatol. 1999;40:433–5.
65. Huang WS, Yang TY, Shen WC, et al. Association between *Helicobacter pylori* infection and dementia. J Clin Neurosci. 2014;21:1355–8.
66. Xu Y, Wang Q, Liu Y, et al. Is *Helicobacter pylori* infection a critical risk factor for vascular dementia? Int J Neurosci. 2016;126:899–903.
67. Uberti AF, Callai-Silva N, Grahl MVC, et al. *Helicobacter pylori* urease: potential contributions to Alzheimer's disease. Int J Mol Sci. 2022;23:3091.
68. Hsieh MC, Wang SS, Hsieh YT, et al. *Helicobacter pylori* infection associated with high HbA1c and type 2 diabetes. Eur J Clin Invest. 2013;43:949–56.
69. Wang F, Fu Y, Lv Z. Association of *Helicobacter pylori* infection with diabetic complications: a meta-analysis. Endocr Res. 2014;39:7–12.
70. Ponzetto A, Cardaropoli S, Piccoli E, et al. Pre-eclampsia is associated with *Helicobacter pylori* seropositivity in Italy. J Hypertens. 2006;24:2445–9.
71. Cardaropoli S, Rolfo A, Todros T. *Helicobacter pylori* and pregnancy-related disorders. World J Gastroenterol. 2014;20:654–64.
72. Eslick GD, Yan P, Xia HH, et al. Foetal intrauterine growth restrictions with *Helicobacter pylori* infection. Aliment Pharmacol Ther. 2002;16:1677–82.
73. Koçak I, Akcan Y, Ustün C, et al. *Helicobacter pylori* seropositivity in patients with hyperemesis gravidarum. Int J Gynaecol Obstet. 1999;66:251–4.

Part II

Diagnosis of *Helicobacter pylori*

Indications for *Helicobacter pylori* Testing During Esophagogastroduodenoscopy

3

Gwang Ha Kim

Highlights

What is known
• In clinical practice, the decision to perform *H. pylori* testing is based on clinical information such as symptoms, medication history, laboratory data, and endoscopic findings. • In cases of health checkup endoscopy, endoscopists often perform endoscopy without any clinical information.
Future research questions
• Research on endoscopic findings that are highly suspicious of aspirin/NSAID use and indicative of long-term PPI is required. • Known endoscopic findings for current *H. pylori* infection should be validated.

3.1 Introduction

Helicobacter pylori (*H. pylori*) infection is the most prevalent chronic bacterial infection in the world and is associated with chronic gastritis such as atrophic gastritis and intestinal metaplasia, peptic ulcer disease, gastric adenocarcinoma, and gastric mucosa-associated lymphoid tissue (MALT) lymphoma [1]. When to diagnose and treat *H. pylori* infection were already described in the previous chapter, and endoscopic diagnosis of *H. pylori* infection will be dealt in the next chapter. Prior to performing esophagogastroduodenoscopy (EGD), endoscopists are aware of patient information such as symptoms, medical history, laboratory data (e.g., dyspepsia, heartburn, use of nonsteroidal anti-inflammatory drugs [NSAIDs], or thrombocytopenia), and previous history of gastric lesions, such as peptic ulcer or gastric epithelial neoplasia, and, as such, perform invasive *H. pylori* tests (rapid urease test, histopathological examination, or culture) during endoscopy. However, in cases of health checkup endoscopy, endoscopists often perform EGD without any clinical information. Accordingly, this chapter discusses when to perform *H. pylori* tests during EGD based on endoscopic findings alone.

3.2 Main Body

3.2.1 Gastric Lesions

To perform endoscopic biopsies for the diagnosis of *H. pylori* infection, it is important to know which gastric diseases require *H. pylori* eradication. According to recent evidence-based guidelines for the treatment of *H. pylori* infection, indications of *H. pylori* eradication are peptic ulcer disease, MALT lymphoma, after endoscopic resection of early gastric cancer (EGC) and gastric adenoma, family history of gastric

G. H. Kim (✉)
Pusan National University School of Medicine, Busan, Korea
e-mail: doc0224@pusan.ac.kr

B.-W. Kim (ed.), *Helicobacter Infection in Clinical Practice*,
https://doi.org/10.1007/978-981-97-8397-7_3

cancer, functional dyspepsia, long-term low-dose aspirin user with a history of peptic ulcer, idiopathic thrombocytopenic purpura, iron-deficiency anemia, and atrophic gastritis/intestinal metaplasia [2]. Among these indications, peptic ulcer diseases, MALT lymphoma, EGC, gastric adenoma, and atrophic gastritis/intestinal metaplasia can be detected during endoscopy. Therefore, when these lesions are detected, *H. pylori* tests are recommended during endoscopy.

In addition, recent developments in endoscopic technology such as high-definition endoscopy and image-enhanced endoscopy, including narrow-band imaging, allow detailed observation of the gastric mucosa by endoscopists. Recently, the Kyoto classification of gastritis, a grading system for endoscopic gastritis to predict the presence or absence of *H. pylori* infection, was introduced [3]. This classification will be described in detail in the next chapter. In brief, characteristic endoscopic findings of *H. pylori*-uninfected individuals include a regular arrangement of collecting venules, fundic gland polyps, red streaks, etc. In contrast, diffuse redness, spotty redness, atrophy, intestinal metaplasia, enlarged or tortuous folds, mucosal swelling, sticky mucus, nodularity, foveolar hyperplastic polyps, and xanthoma are characteristic endoscopic findings in current *H. pylori*-infected individuals. Among these endoscopic findings, diffuse redness, mucosal swelling, enlarged or tortuous folds, sticky mucus, and nodularity can be improved after *H. pylori* eradication. When any one of these five findings are observed during endoscopy, the sensitivity and specificity for predicting current *H. pylori* infection are 92% and 90%, respectively [4]. In previously *H. pylori*-infected persons, patchy redness and maplike redness can be observed. Therefore, considering the strong association between *H. pylori* infection and gastric cancer, and the clear evidence of the reduction in gastric cancer after *H. pylori* eradication [5, 6], *H. pylori* tests are recommended when endoscopic findings suggestive of current *H. pylori* infection are observed.

Both *H. pylori* infection and NSAIDs use are independent risk factors for peptic ulcer disease development. In particular, the use of either aspirin or NSAIDs also increases the risk of peptic ulcer disease and its related complications in *H. pylori*-infected patients [1]. Therefore, it is important to check whether *H. pylori* infection is present during endoscopy to predict and prevent complications associated with aspirin/NSAID use. Although these endoscopic findings are not specific for aspirin/NSAID use, multiple petechial hemorrhages and small red or black erosions are usually observed mainly in the gastric antrum and fundus during endoscopy [7]. Therefore, when these endoscopic findings are seen with or without gastric ulcers, *H. pylori* tests are recommended, especially when the above-mentioned endoscopic findings suggestive of current *H. pylori* infection are observed.

3.2.2 Non-gastric Lesions

Patterns of *H. pylori* colonization and associated gastritis depend on the level of gastric acid output. In patients with normal-to-increased acid output, bacterial colonization and gastritis are predominantly confined to the gastric antrum. In cases with decreased acid output, bacterial colonization and gastritis also affect the gastric body, leading to corpus-predominant pangastritis [8]. Proton-pump inhibitor (PPI) use suppresses gastric acid secretion, resulting in persistent hypergastrinemia. In particular, serum gastrin levels are higher (approximately 1.5-fold) in *H. pylori*-infected patients than in *H. pylori*-uninfected patients. *H. pylori*-infected PPI users exhibit ~2.5 times more enterochromaffin-like cell hyperplasia in the gastric corpus than *H. pylori*-uninfected PPI users [9]. Long-term PPI treatment is associated with a shift from antrum-predominant gastritis to corpus-predominant gastritis and increased corpus atrophic gastritis (odds ratio 11.5) in *H. pylori*-infected patients [9]. Furthermore, *H. pylori* eradication improves gastritis in long-term PPI users irrespective of PPI continuation [10, 11].

Therefore, in patients who require long-term PPI use, such as those with severe reflux esophagitis (LA C, D) and Barrett's esophagus (in Western countries, not yet in Asia), testing for *H. pylori* can be considered, despite the negative association between *H. pylori* infection and gastroesophageal reflux disease and Barrett's esophagus.

Endoscopic findings in the stomach, such as multiple petechial hemorrhages and small red erosions with or without ulcers, are also detected in the first and second portions of the duodenum in aspirin/NSAID users. These endoscopic findings in the stomach and duodenum are simultaneously observed in most cases. As mentioned above, an approach that considers endoscopic findings suggestive of current *H. pylori* infection in the stomach would be ideal for determining whether to perform *H. pylori* tests in patients with multiple petechial hemorrhages and small red erosions in the duodenum.

3.3 Conclusions

In clinical practice, the decision to perform *H. pylori* testing is based on clinical information such as symptoms, medication history, laboratory data, and endoscopic findings. I suggest situations when to perform *H. pylori* tests based on the endoscopic findings alone (Table 3.1). Indications for *H. pylori* eradication, such as peptic ulcer disease, gastric MALT lymphoma, and gastric epithelial neoplasia, are diagnosed only by endoscopy and endoscopic biopsy. Therefore, in cases in which these lesions are detected during EGD, a rapid urease test, histopathological examination, or culture for the diagnosis of *H. pylori* infection should be performed. Aspirin/NSAID use and long-term PPI use, which are indications for *H. pylori* infection, can be identified by a medical chart review. However, several endoscopic findings that are highly suspicious of aspirin/NSAID use and indicative of long-term PPI use can be found during EGD. Therefore, based on these endoscopic findings, *H. pylori* tests could be also consid-

Table 3.1 Recommendation for *H. pylori* tests based on the endoscopic findings

Strong recommendation	• Peptic ulcer diseases • Gastric MALT lymphoma • Gastric epithelial neoplasia (cancer and adenoma)
Suggestive recommendation	• Highly suggestive findings of current *H. pylori* infection – Diffuse redness – Mucosal swelling – Enlarged or tortuous folds – Sticky mucus – Nodularity – Spotty redness – Atrophy – Intestinal metaplasia – Foveolar hyperplastic polyps – Xanthoma
Probable recommendation[a]	• Findings suggestive of aspirin/NSAID use (stomach and duodenum) – Multiple petechial hemorrhages – Small red or black erosions • Findings indicative of long-term PPI use (esophagus) – Severe reflux esophagitis (LA C, D) – Barrett's esophagus

[a]An approach considering endoscopic findings suggestive of current *H. pylori* infection is recommended

ered. Even in these situations, it is reasonable to perform *H. pylori* tests selectively, considering the endoscopic findings of current *H. pylori* infection.

References

1. Malfertheiner P, Megraud F, Rokkas T, et al. Management of Helicobacter pylori infection: the Maastricht VI/Florence consensus report. Gut. 2022;
2. Jung HK, Kang SJ, Lee YC, et al. Evidence-based guidelines for the treatment of Helicobacter pylori infection in Korea 2020. Gut Liver. 2021;15:168–95.
3. Haruma K, Kato M, Inoue K, Murakami M, Kamada T, editors. Kyoto classification of gastritis. Tokyo: Nihon Medical Center; 2017.
4. Otani K, Watanabe T, Kosaka S, et al. Utility of Kyoto classification of gastritis in subjects with a high-negative titer of anti-Helicobacter pylori antibody during a medical check-up. J Clin Biochem Nutr. 2020;67:317–22.
5. Huang JQ, Sridhar S, Chen Y, Hunt RH. Meta-analysis of the relationship between Helicobacter pylori seropositivity and gastric cancer. Gastroenterology. 1998;114:1169–79.
6. Lee YC, Chiang TH, Chou CK, et al. Association between Helicobacter pylori eradication and gastric cancer incidence: a systematic review and meta-analysis. Gastroenterology. 2016;150:1113–24.
7. Sugawa C, Walt AJ, Lucas CE. Differential topography of acute erosive gastritis due to trauma or sepsis, ethanol, and aspirin. Gastroenterology. 1973;19:127–30.
8. Malfertheiner P, Megraud F, O'Morain CA, et al. Management of Helicobacter pylori infection-the Maastricht V/Florence Consensus Report. Gut. 2017;66:6–30.
9. Lundell L, Vieth M, Gibson F, Nagy P, Kahrilas PJ. Systematic review: the effects of long-term proton pump inhibitor use on serum gastrin levels and gastric histology. Aliment Pharmacol Ther. 2015;42:649–63.
10. Kuipers EJ, Nelis GF, Klinkenberg-Knol EC, et al. Cure of Helicobacter pylori infection in patients with reflux oesophagitis treated with long term omeprazole reverses gastritis without exacerbation of reflux disease: results of a randomised controlled trial. Gut. 2004;53:12–20.
11. Schenk BE, Kuipers EJ, Nelis GF, et al. Effect of Helicobacter pylori eradication on chronic gastritis during omeprazole therapy. Gut. 2000;46:615–21.

Endoscopic Diagnosis for *Helicobacter pylori* Infection

4

Sun-Young Lee

Highlights

What is Known

- Diffuse or spotty redness in the corpus, enlarged gastric folds, mucosal swelling, sticky mucus, and antral nodularity suggest a current *H. pylori* infection. Conversely, regular arrangement of collecting venules in the angle indicates a stomach without *H. pylori*.
- To avoid false–negative *H. pylori* invasive test findings, it is important to choose appropriate sites for biopsy during gastroscopy. Obtain biopsies from the hyperemic mucosa in the greater curvature of the upper body, which shows diffuse redness.
- Invasive tests should not be used for eradication confirmation.

Future Research Questions

- Beyond *H. pylori* infection, what are the causes and proportions of redness, mucosal swelling, and enlarged gastric folds?
- What are the discrimination points between the spotty redness induced by *H. pylori* infection and those induced by potassium-competitive acid blockers?
- How often does a long-term intake of acid suppressants induce spotty redness?
- Why does the portal hypertensive gastropathy induce spotty redness? What is different from *H. pylori*-infected stomachs?
- What is the next step when a false-negative or a false-positive invasive test finding is suspicious in daily clinical practice? Which *H. pylori* test is most cost-effective?

S.-Y. Lee (✉)
Department of Internal Medicine, Konkuk University School of Medicine, Seoul, Korea
e-mail: sunyoung@kuh.ac.kr

4.1 Introduction

In areas endemic to *Helicobacter pylori* (*H. pylori*) infection, a national screening program for gastric cancer is often performed using endoscopic examination. During endoscopic screening, it is important to obtain gastric biopsies for *H. pylori* tests when the endoscopic findings suggest an infection. Herein, the endoscopic findings suggesting ongoing *H. pylori* infection are summarized. Furthermore, the types and limitations of invasive methods used for *H. pylori* confirmation are described.

4.2 Main Body

4.2.1 Endoscopic Findings Suggesting Ongoing *H. pylori* Infection

4.2.1.1 Diffuse Redness in the Corpus

In *H. pylori*-infected stomachs, diffuse redness of the corpus is one of the common gastroscopy findings (Fig. 4.1). Diffuse redness is associated with inflammatory cell infiltration [1]. This redness can be overlooked when there is progression to gastric corpus atrophy (Fig. 4.2). In the absence of atrophy, it is easier to detect diffuse redness (Fig. 4.3).

Diffuse redness is inversely correlated with the regular arrangement of collecting venules

B.-W. Kim (ed.), *Helicobacter Infection in Clinical Practice*,
https://doi.org/10.1007/978-981-97-8397-7_4

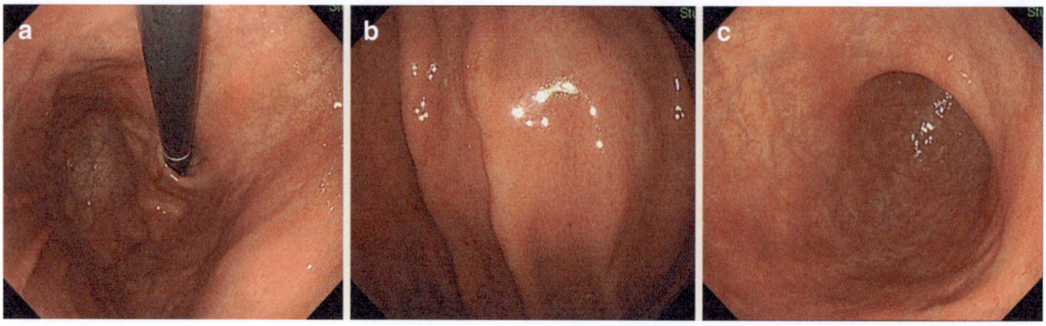

Fig. 4.1 Gastroscopic images showing diffuse redness in an infected stomach. (**a**) Diffuse redness and atrophy are observed in the retroflexed view of the cardia and fundus. (**b**) Diffuse redness is observed in the gastric rugae. (**c**) Extensive atrophy is observed in the lower body and antrum

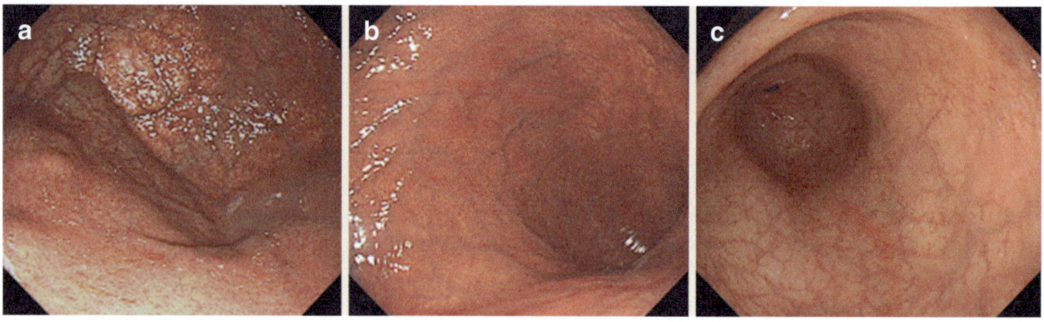

Fig. 4.2 Diffuse redness in an infected stomach with open-type atrophic gastritis without gastric rugae. (**a**) Diffuse redness is observed in the retroflexed view of the fundus. (**b**) Diffuse redness is masked by extensive atrophy. (**c**) Extensive atrophy is observed from the antrum to the corpus without an atrophic border. The findings are consistent with open-type pangastritis

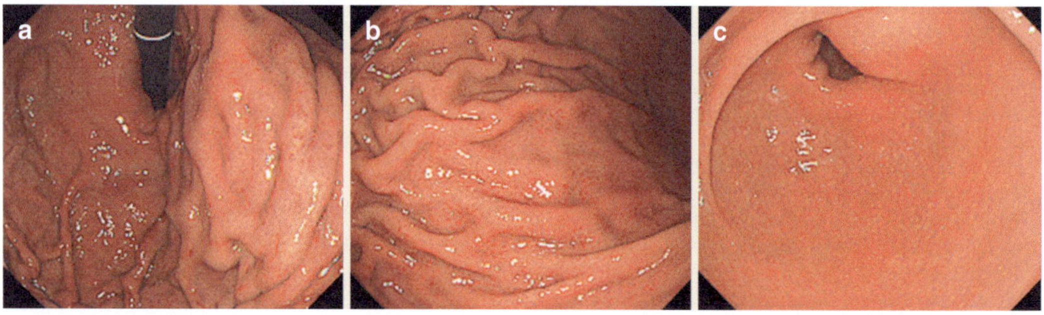

Fig. 4.3 Diffuse redness in an infected stomach with no atrophy. (**a**) Diffuse redness is observed in the retroflexed view of the cardia. (**b**) Diffuse redness and swelling are observed in the gastric rugae. (**c**) There is no atrophy in the antrum

(RAC) [2]. In a recent Korean study, *H. pylori* infection rates in individuals with RAC, mosaic-like gastric mucosal surfaces, diffuse redness, and atypical findings were 9.4%, 87.7%, 98.1%, and 90.9%, respectively [3].

4.2.1.2 Spotty Redness in the Corpus

Spotty redness is easier to detect than diffuse redness; hence, it is often used as an indicator after eradication therapy [4]. It regresses within a few months after eradication [5]. Moreover, it

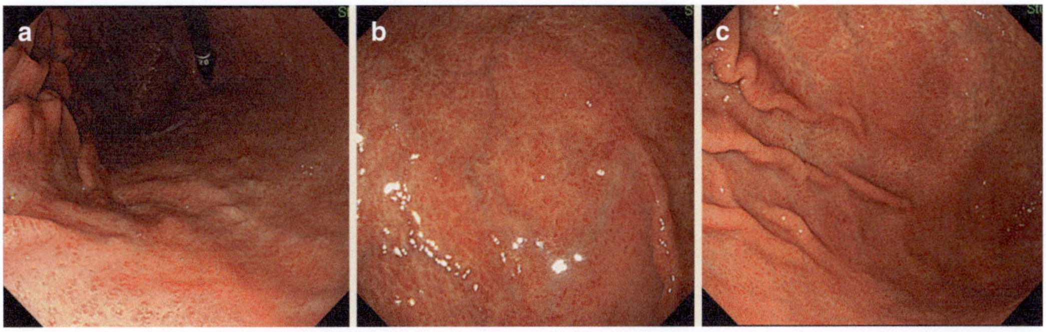

Fig. 4.4 An infected stomach with spotty redness. (**a**) In the retroflexed view, multiple tiny hemorrhagic spots are observed. (**b**) Spotty redness is observed in the fundus. (**c**) Spotty redness is observed in the greater curvature of the body

reappears when there is reinfection or recrudescence of *H. pylori* [6]. Similar to diffuse redness, spotty redness appears in the corpus of an *H. pylori*-infected stomach (Fig. 4.4). Since it is associated with the distribution of neutrophils in the gastric mucosa, spotty redness is included within the diffuse redness score in the Kyoto classification: (1) atrophy scores from 0 to 2, (2) intestinal metaplasia scores from 0 to 2, (3) hypertrophic rugae scores from 0 to 1, (4) nodular gastritis scores from 0 to 1, and (5) diffuse redness (including spotty redness) scores from 0 to 2 [7].

4.2.1.3 Enlarged Gastric Folds (Hypertrophic Rugae)

Hypertrophic gastric folds ≥5 mm are often found with sticky exudate and mucosal swelling in *H. pylori*-infected stomachs (Fig. 4.5). In a recent Japanese study, the odds ratios (OR) for detecting *H. pylori* infection were as high as 8.6 in the presence of enlarged folds [8]. Diffuse redness showed the highest odds ratio (OR = 26.8), followed by mucosal swelling (OR = 13.3), sticky mucus (OR = 10.2), and enlarged folds (OR = 8.6). Improvement of enlarged folds should be monitored after successful *H. pylori* eradication because some Borrmann-type IV gastric cancers also exhibit hypertrophic rugae [9]. It is important to discriminate between hypertrophic gastric folds induced by *H. pylori*-associated inflammation and gastric cancer with submucosal invasion [10].

4.2.1.4 Nodular Gastritis in the Antrum

Chicken skin-like nodularity in the antrum is another endoscopic finding suggestive of current *H. pylori* infection [1]. Nodular gastritis is observed as multiple, regular-sized, subepithelial nodules in the antrum (Fig. 4.6). It is more common in the younger population aged below 20 years, in whom atrophy is rare [11]. Gastric biopsies from the nodules often show *H. pylori* infection with lymphoid follicles and lymphoid cell aggregates [12]. In East Asia, nodular gastritis is regarded as a precancerous condition of diffuse-type gastric cancer [13]. Furthermore, some gastric mucosa-associated lymphoid tissue lymphomas induced by *H. pylori* or *Helicobacter suis* infection can be observed as nodular gastritis [14]. Nonetheless, it is regarded as an acute immune response outside East Asia [15].

In a previous study, we found that in patients with persistent infection, small-granular-type nodular gastritis often progresses to atrophy, whereas large-nodular-type nodular gastritis often progresses to intestinal metaplasia [16]. A salt-and-pepper appearance is observed after the regression of small-granular-type, nodular gastritis, which consists of 1–2 mm-sized nodules. In the large-nodular-type, whitish plaques and diffuse irregular elevations are observed after the regression of 3–4 mm-sized nodules.

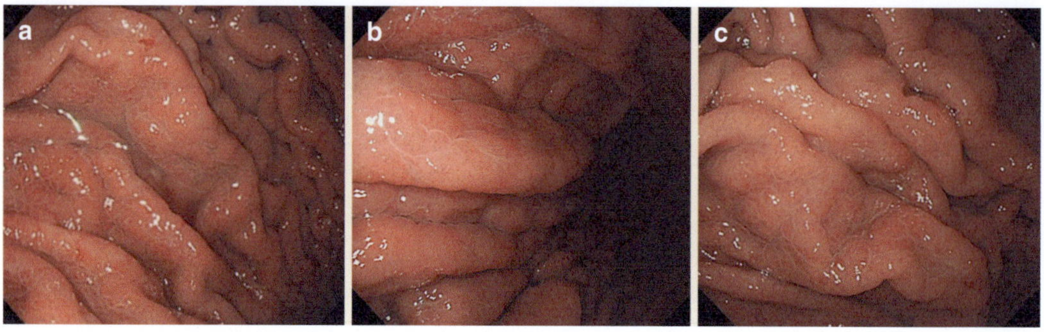

Fig. 4.5 An infected stomach with thick rugae. (**a**) Hypertrophic gastric folds and spotty redness are observed in the upper body. (**b**) Edematous change is observed in the thick rugae. (**c**) Diffuse redness and mucosal swelling are observed in the rugae of the lower body

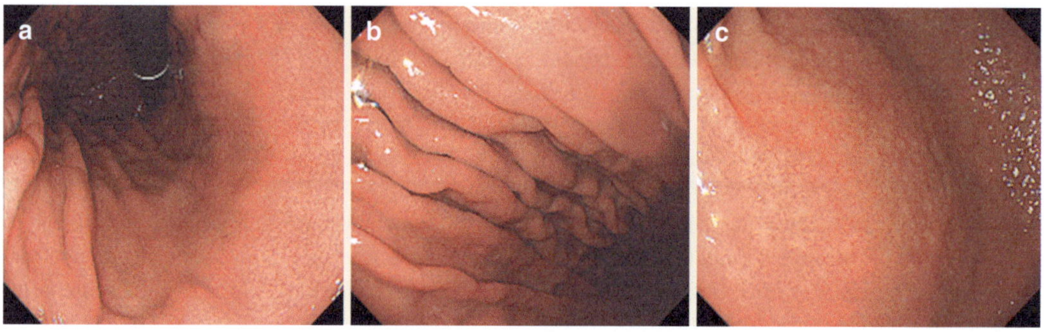

Fig. 4.6 An infected stomach with nodular gastritis. (**a**) Diffuse redness is observed in the retroflexed view of the corpus. (**b**) Diffuse redness is observed in the thick rugae. (**c**) Chicken skin-like nodularity is observed in the antrum

4.2.1.5 Focal Lesions (Gastric Xanthomas, Hyperplastic Polyps, Ulcers, and Tumors)

In addition to gastroscopy findings of the background gastric mucosa, several focal lesions are often found in the infected stomachs. Gastric xanthoma is a yellowish plaque found in stomachs with chronic or past *H. pylori* infection (Fig. 4.7). It is also called as gastric xanthelasma. Both xanthoma and xanthelasma indicate increased risks of intestinal-type gastric cancer [17].

A considerable number of hyperplastic polyps arise from inflammation caused by *H. pylori* [18]. Small-sized hyperplastic polyps may regress after successful eradication. Conversely, persistent infections may increase the size and number of hyperplastic polyps [19]. Most hyperplastic polyps induced by *H. pylori* infection are red in color (Fig. 4.8).

Gastric ulcers and neoplasms may also occur in infected stomachs. In the presence of advanced atrophy or intestinal metaplasia, slow-growing intestinal-type gastric cancer or gastric adenoma may occur [20]. Conversely, diffuse-type gastric cancer may occur without atrophy or intestinal metaplasia [21]. Enlarged gastric folds are often observed in patients with diffuse-type gastric cancer [22].

4.2.2 How to Increase the Efficacy of Invasive *H. pylori* Tests

4.2.2.1 Avoid *H. pylori* Tests in Noninfected Stomachs

It is important to discriminate *H. pylori*-infected stomachs from naive and previously infected stomachs [23]. Biopsy is not required for *H. pylori* tests if RAC is observed in the angle and in

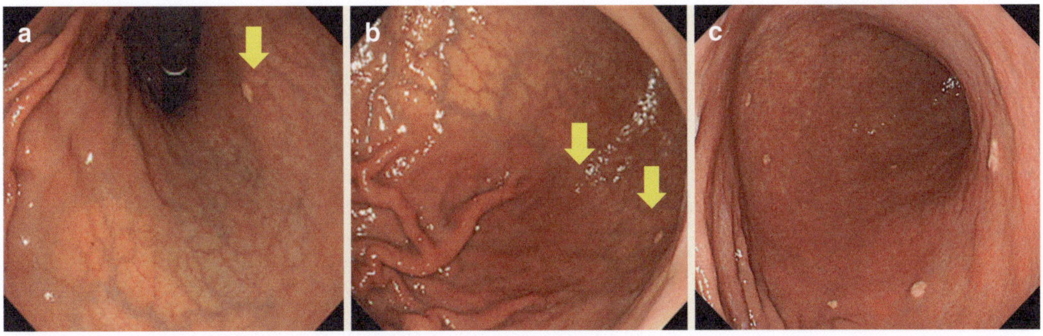

Fig. 4.7 An infected stomach with xanthomas. (**a**) In the retroflexed view, open-type atrophy and xanthoma (arrow) are observed. (**b**) Edematous gastric folds are observed with xanthomas (arrows) behind the atrophic border. (**c**) Several tiny xanthomas are observed from the lower body to the proximal antrum

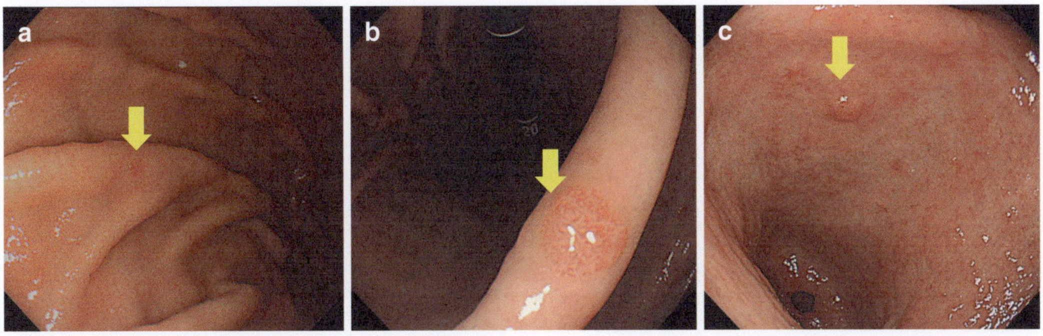

Fig. 4.8 An infected stomach with hyperplastic polyps. (**a**) A tiny hyperemic polyp (arrow) is observed in the gastric rugae. (**b**) Another hyperemic polyp (arrow) is observed in the angle. (**c**) Multiple tiny polypoid lesions are scattered in the antrum. Biopsy finding of the polyp (arrow) was consistent with hyperplastic polyps

the lesser curvature of the corpus. RAC is a valuable endoscopic feature of uninfected status with high accuracy [24]. In addition to RAC, fundic gland polyps and linear red streaks suggest an uninfected status [2]. Contrarily, maplike redness suggests past infection; hence, biopsy is not required in atrophic or metaplastic conditions showing geographic, hyperemic, and depressed erosions [8]. Furthermore, multiple white and flat elevated lesions, as well as gastric black spots, also suggest past infection [25].

Conversely, *H. pylori* invasive test is needed if thick rugae, mucosal swelling, or diffuse redness are observed in the background gastric mucosa. These endoscopic findings should be discriminated from hematins and antral erosions because they indicate a hyperchlorhydric gastric environment regardless of *H. pylori* infection status [26].

4.2.2.2 Choose Appropriate Sites for Biopsy

The location, size, and quality of the sample are important to avoid sampling errors (Table 4.1). A sufficient amount of mucosa should be obtained from the greater curvature of the upper body (Table 4.2). Because of the uneven distribution of *H. pylori*, several biopsies from various sites can improve accuracy [27]. Hyperemic gastric mucosa without atrophy, intestinal metaplasia, ulcer, or bleeding should be selected for biopsy.

4.2.2.3 Do Not Use Invasive Tests for Eradication Confirmation

Recent use of antibiotics, PPIs, P-CABs, or bismuth compounds within 4 weeks decreases the diagnostic accuracy rate; hence, a urea breath test or monoclonal stool antigen test is needed to confirm the eradication. Histamine-2 receptor antag-

Table 4.1 Common causes of false test findings

Invasive *H. pylori* tests	False-negative test findings	False-positive test findings
Rapid urease test	• Recent use of antibiotics, bismuth compounds, or acid suppressants • Sampling at *H. pylori*-scarce areas[a]	• Urease-producing bacteria
Histology	• Lack of special staining for *H. pylori* (*i.e.*, Giemsa stain) • Recent use of antibiotics, bismuth compounds, or acid suppressants • Sampling at *H. pylori*-scarce areas[a]	• Non-*H. pylori Helicobacter* • *Campylobacter*
Polymerase chain reaction	• Sampling at *H. pylori*-scarce areas[a]	• DNA from dead *H. pylori*
Culture	• Slow and difficult growth of *H. pylori* • Sampling at *H. pylori*-scarce areas[a]	• Contamination

[a]*H. pylori*-scarce areas including gastric neoplasm, intestinal metaplasia, atrophy, and hemorrhage

Table 4.2 Recent studies on appropriate biopsy sites for *H. pylori* tests

Recommended sites for gastric biopsies during endoscopic examination	Study subjects	References
Corpus	310 *H. pylori*-infected patients	Cho et al. (2017) [29]
Corpus	164 dyspepsia patients	Lan et al. (2012) [30]
Corpus	70 gastric ulcer patients	Bermejo et al. (2002) [31]
Greater curvature of the corpus	91 patients with extensive atrophy	Lee et al. (2012) [32]
Greater curvature of the upper body	194 gastric cancer patients	Kim et al. (2009) [33]
Proximal stomach and angle	103 gastric ulcer patients	Ho et al. (2004) [34]
Anastomotic site and mid-to-high body	42 patients after gastrectomy	Kubota et al. (2002) [35]

onists (H2-blockers) do not affect the accuracy of *H. pylori* invasive tests, whereas PPIs and P-CABs often lead to false-negative *H. pylori* test findings [28]. Therefore, H2-blockers are recommended instead of PPIs or P-CABs before the *H. pylori* tests.

4.3 Conclusions

To avoid false-negative *H. pylori* invasive test findings, it is important to identify the *H. pylori*-infected stomachs. Diffuse or spotty redness in the corpus, mucosal swelling, enlarged folds, sticky mucus, and antral nodularity suggest a current infection. Biopsy is not required for *H. pylori* tests if RAC is observed in the angle and in the lesser curvature of the corpus. Furthermore, obtaining biopsies from the hyperemic gastric mucosa in the greater curvature of the upper body is important to avoid sampling error. Lastly, invasive tests should not be used to confirm the eradication.

References

1. Toyoshima O, Nishizawa T, Yoshida S, Matsuno T, Odawara N, Toyoshima A, Sakitani K, Watanabe H, Fujishiro M, Suzuki H. Consistency between the endoscopic Kyoto classification and pathological updated Sydney system for gastritis: a cross–sectional study. J Gastroenterol Hepatol. 2022;37:291–300.
2. Ebigbo A, Marienhagen J, Messmann H. Regular arrangement of collecting venules and the Kimura–Takemoto classification for the endoscopic diagnosis of Helicobacter pylori infection: evaluation in a Western setting. Dig Endosc. 2021;33:587–91.

3. Cho JH, Chang YW, Jang JY, Shim JJ, Lee CK, Dong SH, Kim HJ, Kim BH, Lee TH, Cho JY. Close observation of gastric mucosal pattern by standard endoscopy can predict Helicobacter pylori infection status. J Gastroenterol Hepatol. 2013;28:279–84.
4. Kato M, Terao S, Adachi K, Nakajima S, Ando T, Yoshida N, Uedo N, Murakami K, Ohara S, Ito M, Uemura N, Shimbo T, Watanabe H, Kato T, Ida K, Study Group for Establishing Endoscopic Diagnosis of Chronic Gastritis. Changes in endoscopic findings of gastritis after cure of H. pylori infection: multicenter prospective trial. Dig Endosc. 2013;25:264–73.
5. Toyoshima O, Nishizawa T, Sakitani K, Yamakawa T, Takahashi Y, Kinoshita K, Torii A, Yamada A, Suzuki H, Koike K. Helicobacter pylori eradication improved the Kyoto classification score on endoscopy. JGH Open. 2020;4:909–14.
6. Kim YJ, Lee SY, Kim JH, Sung IK, Park HS. Incidence of infection among subjects with Helicobacter pylori seroconversion. Clin Endosc. 2022;55:67–76.
7. Toyoshima O, Nishizawa T, Yoshida S, Sakaguchi Y, Nakai Y, Watanabe H, Suzuki H, Tanikawa C, Matsuda K, Koike K. Endoscopy–based Kyoto classification score of gastritis related to pathological topography of neutrophil activity. World J Gastroenterol. 2020;26:5146–55.
8. Yoshii S, Mabe K, Watano K, Ohno M, Matsumoto M, Ono S, Kudo T, Nojima M, Kato M, Sakamoto N. Validity of endoscopic features for the diagnosis of Helicobacter pylori infection status based on the Kyoto classification of gastritis. Dig Endosc. 2020;32:74–83.
9. Lee SY. Giant rugae in Helicobacter pylori–infected stomachs. Clin Endosc. 2021;54:1–3.
10. Toyoshima O, Yoshida S, Nishizawa T, Toyoshima A, Sakitani K, Matsuno T, Yamada T, Matsuo T, Nakagawa H, Koike K. Enlarged folds on endoscopic gastritis as a predictor for submucosal invasion of gastric cancers. World J Gastrointest Endosc. 2021;13:426–36.
11. Hojo M, Nagahara A, Kudo T, Takeda T, Ikuse T, Matsumoto K, Ueda K, Ueyama H, Matsumoto K, Asaoka D, Shimizu T. Endoscopic findings of Helicobacter pylori gastritis in children and young adults based on the Kyoto classification of gastritis and age–associated changes. JGH Open. 2021;5:1197–202.
12. Kim YJ, Lee SY, Lee SP, Kim JH, Sung IK, Park HS, Shim CS, Han HS. Identification of nodular gastritis among patients diagnosed with lymphofollicular gastritis on a gastric biopsied specimen. Korean J Gastroenterol. 2018;71:143–52.
13. Okamoto K, Kodama M, Mizukami K, Okimoto T, Abe H, Ogawa R, Fukuda K, Matsunari O, Hirashita Y, Wada Y, Fukuda M, Murakami K. Immunohistochemical differences in gastric mucosal damage between nodular and non–nodular gastritis caused by Helicobacter pylori infection. J Clin Biochem Nutr. 2021;69:216–21.
14. Takigawa H, Masaki S, Naito T, Yuge R, Urabe Y, Tanaka S, Sentani K, Matsuo T, Matsuo K, Chayama K, Kitadai Y. Helicobacter suis infection is associated with nodular gastritis–like appearance of gastric mucosa–associated lymphoid tissue lymphoma. Cancer Med. 2019;8:4370–9.
15. Mansilla-Vivar R, Serrano CA, Palma C, Vera M, Hernandez C, Pizarro M, Torres J, Harris PR, Fuentes-López E, Riquelme A, Espino A. High Helicobacter pylori bacterial load and low cytokine expression levels are associated with nodular gastropathy. Dig Dis Sci. 2020;65:565–75.
16. Kim YJ, Lee SY, Yang H, Kim JH, Sung IK, Park HS. Nodular gastritis as a precursor lesion of atrophic and metaplastic gastritis. Korean J Gastroenterol. 2019;73:332–40.
17. Kitamura S, Muguruma N, Okamoto K, Tanahashi T, Fukuya A, Tanaka K, Fujimoto D, Kimura T, Miyamoto H, Bando Y, Takeuchi H, Shiba M, Takayama T. Clinicopathological assessment of gastric xanthoma as potential predictive marker of gastric cancer. Digestion. 2017;96:199–206.
18. Jeong CY, Kim N, Lee HS, Yoon H, Shin CM, Park YS, Kim JW, Lee DH. Risk factors of multiple gastric polyps according to the histologic classification: prospective observational cohort study. Korean J Gastroenterol. 2019;74:17–29.
19. Nam SY, Lee SW, Jeon SW, Kwon YH, Lee HS. Helicobacter pylori eradication regressed gastric hyperplastic polyp: a randomized controlled trial. Dig Dis Sci. 2020;65:3652–9.
20. Lee HH, Lee SY, Yoon HY, Hong SN, Kim JH, Sung IK, Park HS, Shim CS, Han HS. Change of mucosal color in early gastric cancer. J Dig Dis. 2012;13:510–6.
21. Toyoshima O, Nishizawa T. Kyoto classification of gastritis: advances and future perspectives in endoscopic diagnosis of gastritis. World J Gastroenterol. 2022;28:6078–89.
22. Toyoshima O, Nishizawa T, Yoshida S, Aoki T, Nagura F, Sakitani K, Tsuji Y, Nakagawa H, Suzuki H, Koike K. Comparison of endoscopic gastritis based on Kyoto classification between diffuse and intestinal gastric cancer. World J Gastrointest Endosc. 2021;13:125–36.
23. Lee SY. Comparison of gastric cancer screening strategies between Helicobacter pylori-infected, -eradicated, and -naive individuals. Korean J Helicobacter Up Gastrointest Res. 2022;22:313–6.
24. Li L, Jing J, Gao H, Zhang C, Lou H, Pan W. Regular arrangement of collecting venules under endoscopy for predicting a Helicobacter pylori–negative stomach: a systematic review and meta–analysis. Gastroenterol Hepatol. 2021;44:286–92.
25. Hatano Y, Haruma K, Kamada T, Shiotani A, Takahari K, Matsumoto M, Uchida O. Factors associated with gastric black spot, white flat elevated mucosa, and cobblestone–like mucosa: a cross–sectional study. Digestion. 2018;98:185–93.

26. Hatta W, Iijima K, Koike T, Kondo Y, Ara N, Asanuma K, Uno K, Asano N, Imatani A, Shimosegawa T. Endoscopic findings for predicting gastric acid secretion status. Dig Endosc. 2015;27: 582–9.
27. Moon SW, Kim TH, Kim HS, Ju JH, Ahn YJ, Jang HJ, Shim SG, Kim HJ, Jung WT, Lee OJ. United rapid urease test is superior than separate test in detecting Helicobacter pylori at the gastric antrum and body specimens. Clin Endosc. 2012;45:392–6.
28. Graham DY, Opekun AR, Jogi M, Yamaoka Y, Lu H, Reddy R, El-Zimaity HM. False negative urea breath tests with H2–receptor antagonists: interactions between Helicobacter pylori density and pH. Helicobacter. 2004;9:17–27.
29. Cho JH, Jeon SR, Kim HG, Jin SY, Park S. Factors for improving the diagnostic efficiency of the rapid urease test from the gastric corpus. Scand J Gastroenterol. 2017;52:1320–5.
30. Lan HC, Chen TS, Li AF, Chang FY, Lin HC. Additional corpus biopsy enhances the detection of Helicobacter pylori infection in a background of gastritis with atrophy. BMC Gastroenterol. 2012;12:182.
31. Bermejo F, Boixeda D, Gisbert JP, Defarges V, Sanz JM, Redondo C, Martiní de Argila C, García Plaza A. Rapid urease test utility for Helicobacter pylori infection diagnosis in gastric ulcer disease. Hepatogastroenterology. 2002;49:572–5.
32. Lee JH, Park YS, Choi KS, Kim DH, Choi KD, Song HJ, Lee GH, Jang SJ, Jung HY, Kim JH. Optimal biopsy site for Helicobacter pylori detection during endoscopic mucosectomy in patients with extensive gastric atrophy. Helicobacter. 2012;17:405–10.
33. Kim CG, Choi IJ, Lee JY, Cho SJ, Nam BH, Kook MC, Hong EK, Kim YW. Biopsy site for detecting Helicobacter pylori infection in patients with gastric cancer. J Gastroenterol Hepatol. 2009;24:469–74.
34. Ho CY, Chen TS, Chang FY, Lee SD. Rapid urease test from non–ulcer part of stomach is superior to histology from ulcer in detection of Helicobacter pylori infection in patients with gastric ulcer. Hepatogastroenterology. 2004;51:1877–80.
35. Kubota K, Shimoyama S, Shimizu N, Noguchi C, Mafune K, Kaminishi M, Tange T. Studies of 13C–urea breath test for diagnosis of Helicobacter pylori infection in patients after partial gastrectomy. Digestion. 2002;65:82–6.

Rapid Urease Test

5

Dae Young Cheung and Soo-Heon Park

Highlights

What is Known
• The RUT is easy and cost-effective method for *H. pylori* detection and test result is usually obtained within 30 min.
• The RUT test has a sensitivity ranging from 93% to 97% and a specificity of approximately 98%.
Future Research Questions
• How can we minimize the occurrence of false positives and false negatives?
• How can we objectively determine RUT results through color discrimination?

5.1 Introduction

The rapid urease Test (RUT) is an invasive diagnostic method that employs tissue acquired through an endoscopic procedure to verify *Helicobacter pylori* infection by detecting the presence of urease elements. As fresh specimens obtained through forceps biopsy are necessary, the examination is conducted by the physician in an endoscopy room.

The standard test method involves an agar gel containing urea and a pH indicator. When gastric mucosal tissue infected with *H. pylori* is inoculated onto the gel, urease of *H. pylori* hydrolyzes urea, leading to the production of ammonia and carbon dioxide. As the pH of the gel increases, the pH indicator triggers a color change in the test media.

The advantages of RUT include its ease to use, cost-effectiveness, and rapid results, usually obtainable within 30 min. These characteristics make RUT suitable for both the initial diagnosis of *H. pylori* infection and follow-up assessment after treatment.

The RUT test demonstrates a sensitivity ranging from 93% to 97% and a specificity of approximately 98%. Test accuracy can be affected by various biological conditions. Atrophic changes in the gastric mucosa or the use of gastric acid secretion inhibitors such as PPIs may inhibit the survival and proliferation of *H. pylori*, leading to a false negative RUT result. Conversely, a weakened stomach acidity may allow urease-producing microbes, such as *Proteus*, to proliferate and result in a false positive.

5.2 Main Body

5.2.1 How the Rapid Urease Test Works

The RUT, also known as the CLO test (*Campylobacter*-like organism test), is a method

D. Y. Cheung · S.-H. Park (✉)
Department of Internal Medicine, Yeouido St. Mary's Hospital, The Catholic University of Korea, Seoul, Korea
e-mail: adagio@catholic.ac.kr

B.-W. Kim (ed.), *Helicobacter Infection in Clinical Practice*,
https://doi.org/10.1007/978-981-97-8397-7_5

used to diagnose the presence of *H. pylori* in the stomach [1]. This diagnostic approach relies on the property of *H. pylori* to produce the enzyme urease, which hydrolyzes urea, breaking it down into ammonia and carbon dioxide. The fundamental process of this test involves collecting a mucosa sample using biopsy forceps during an endoscopy. A piece of mucosal tissue presumed to contain *H. pylori* is then placed in a test medium or strip containing urea and a pH indicator.

In the case of *H. pylori* infection, the enzyme urease breaks down urea in the medium, generating ammonia. This increase in ammonia raises the surrounding pH, causing the pH indicator to change the color of medium. Phenol red is commonly used as the pH indicator, and a positive *H. pylori* infection is indicated by the color change from yellow to pink or magenta. This transformation occurs relatively quickly, typically within minutes or hours.

5.2.2 Accuracy of the RUT

Both the CLO and the ^{13}C-UBT demonstrated higher accuracy in detecting of *H. pylori*. There is no definitive consensus on the gold standard for diagnosing *H. pylori* infection. While a culture-based test is considered the most accurate from a microbiological perspective, in real clinical practice, histological testing with Giemsa staining is recognized as the standard for diagnosis.

In comparison to the histological examination, both the RUT and the ^{13}C-UBT exhibited acceptable sensitivity (93.6% and 96.2%) and accuracy (93.8% and 93.8%). The specificity and positive predictive value for the RUT were superior to the ^{13}C-UBT (94.3% vs. 88.6%, 97.3% vs. 94.9%, respectively) [2].

5.2.3 *H. pylori* and Urease Activity

Urease activity refers to the ability of *H. pylori* to produce and secrete the enzyme urease, which is crucial for its survival in the acidic environment of the stomach. The enzyme urease breaks down urea within the mucus layer, generating ammonia and carbon dioxide. The ammonia produced by urease plays various roles in the survival and pathogenesis of *H. pylori*. Firstly, it acts as a buffer, defending against the bacteriostatic or sterilizing effects of gastric acid, enabling *H. pylori* to stably colonize and proliferate on the mucus layer. Secondly, as an alkaline substance, ammonia may cause direct damage to the gastric epithelium, triggering an inflammatory response that complements inflammation caused by *H. pylori*.

5.2.4 Collection and Preparation for the Test

Mucosal tissue samples are obtained using endoscopic biopsy forceps during esophagogastroduodenoscopy. To enhance the accuracy of the test, it is crucial to choose healthy mucosa where *H. pylori* can effectively colonize and proliferate. Biopsies of both antrum and corpus mucosa, devoid of atrophic or metaplastic changes, are highly recommended as the ideal biopsy targets [3].

The RUT is available in liquid-based, gel-based, or dry-cool (strip) forms, depending on the manufacturer [4]. Therefore, the obtained mucosal tissue sample can be homogenized and prepared as a suspension for a liquid-based test kit. However, gel-based kits are the most commonly used, proven to be more accurate than dry kits, and the biopsied mucosal piece can be used without further processing [5]. In a recent study, mucosal swap preparation also demonstrated proper diagnostic accuracy when compared to tissue-using RUT [6].

Once the tissue sample is prepared, inoculation onto the medium should follow the manufacturer's instructions and explanations. The inoculated medium requires incubation at an appropriate temperature, typically 37 °C, matching body temperature, with an incubation time ranging from 1 to 24 h. Results are determined based on the color change of the medium containing the pH indicator.

5.2.5 Limitation of the Test

The RUT comprises a pH indicator and urea. If conditions unrelated to *H. pylori* that can elevate the pH of the media are present, a false-positive result may occur. For example, excessive salivation or alkaline bile reflux can elevate the pH of the biopsy specimen above normal levels. Medications that inhibit gastric acid secretion or reduce stomach acidity may also lead to false-negative results. In situation where bacteria other than *H. pylori*, which secrete urease, can colonize the stomach, false-positive RUT results may occur. Recognized urease-producing organisms include *Proteus mirabilis, Citrobacter freundii, Klebsiella pneumoniae, Enterobacter cloacae*, and *Staphylococcus aureus* [7].

Another issue is the determination of color change, and if the alteration is not distinct, distinguishing between positive and negative outcomes becomes difficult. In such instances, there is a significant risk of subjective interpretation by the observer. Additionally, there is no standard for the time until the decision is made, which makes it difficult to make a decision after the test. While most benign changes manifest within 30 min to 1 h, there is ongoing debate about whether discoloration after 24 h should be considered benign. Recent research has explored methods to objectively determine results through color discrimination [8].

5.3 Conclusion

The RUT is an easy and cost-effective method for *H. pylori* detection, and the test result is usually obtained within 30 min. The RUT is suitable for both the initial diagnosis of *H. pylori* infection and follow-up assessment after treatment. The RUT test has a sensitivity ranging from 93% to 97% and a specificity of approximately 98%. Various conditions including atrophic change or PPI use can reduce test accuracy.

References

1. Malfertheiner P, Enrique Domínguez-Muñoz J, Heckenmüller H, et al. Modified rapid urease test for detection of Helicobacter pylori infection. Eur J Gastroenterol Hepatol. 1996;8:53–6.
2. Ho AS, Young TH, Shyu RY, et al. The accuracy of the rapid urease test and 13C-urea breath test in the diagnosis of Helicobacter pylori infection. Zhonghua Yi Xue Za Zhi (Taipei). 1996;58:400–6.
3. Parihar V, Holleran G, Hall B, et al. A combined antral and corpus rapid urease testing protocol can increase diagnostic accuracy despite a low prevalence of Helicobacter pylori infection in patients undergoing routine gastroscopy. United Eur Gastroenterol J. 2015;3:432–6.
4. Dolak W, Bilgilier C, Stadlmann A, et al. A multicenter prospective study on the diagnostic performance of a new liquid rapid urease test for the diagnosis of Helicobacter pylori infection. Gut Pathog. 2017;9:78.
5. van Keeken N, van Hattum E, de Boer WA. Validation of a new, commercially available dry rapid urease test for the diagnosis of Helicobacter pylori infection in gastric biopsies. Neth J Med. 2006;64:329–33.
6. Soh H, Kim N, Moon J, et al. Gastric mucosal swab as an alternative to biopsy for Helicobacter pylori detection. Helicobacter. 2023;28:e12980.
7. Osaki T, Mabe K, Hanawa T, et al. Urease-positive bacteria in the stomach induce a false-positive reaction in a urea breath test for diagnosis of Helicobacter pylori infection. J Med Microbiol. 2008;57:814–9.
8. Pih GY, Noh JH, Ahn JY, et al. Objective interpretation of the rapid urease test for Helicobacter pylori infection using colorimetry. J Korean Med Sci. 2022;37:e227.

6 Histopathological Examinations

Sung Hak Lee

Highlights

What is Known
• Histopathology remains a cornerstone in the diagnosis, characterized by its high specificity and ability to provide associated gastric pathology. Techniques like H&E staining, IHC, and special stains like Giemsa have been thoroughly discussed, highlighting their advantages and limitations. • Diverse molecular methods have been evaluated for its high sensitivity and specificity to detect *H. pylori* DNA or RNA even in small amounts. They also provide insight into the genetic makeup of *H. pylori* and are important for understanding antibiotic resistance and guiding treatment methods.
Future Research Questions
• Role of digital pathology and AI to improve accuracy of histopathological diagnosis of *H. pylori*. • Integration of histopathological and molecular data to better understand the progression of *H. pylori*-related diseases and the development of complications.

6.1 Introduction

Tissue examination is currently one of the standard tests for diagnosing *Helicobacter pylori* infection [1]. In addition to diagnosing *H. pylori* infection, tissue examination can provide additional pathological information such as atrophy of the gastric mucosa, inflammation, and intestinal metaplasia. The diagnostic precision of histology is influenced by several factors, such as the location, and number of biopsied samples, staining procedures and quality, the use of proton pump inhibitors, antibiotics, as well as the experience of the pathologists [2].

S. H. Lee (✉)
Department of Hospital Pathology, Seoul St. Mary's Hospital, The Catholic University of Korea, Seoul, Korea
e-mail: hakjjang@catholic.ac.kr

6.2 Main Body

6.2.1 Histologic Methods

The common hematoxylin-eosin (H&E) stain is known to show a sensitivity of 42–99% and a specificity of 87–90% [3, 4]. H&E staining alone may not be enough to find the non-spiral, spherical *H. pylori* due to low sensitivity, so special staining methods like Giemsa, Warthin–Starry, toluidine blue, acridine orange, and Genta should be used in conjunction to increase the diagnostic rate (Fig. 6.1) [1, 3, 5, 6]. By using special staining methods such as Giemsa or Warthin–Starry in addition to H&E staining, the specificity can be increased to 90–100%, so it is recommended to use special staining methods [4]. Among these, the Giemsa staining method is the most widely used because it has a high sensitivity, is cost-effective, and has a high reproducibility [3]. The recently developed modified Giemsa staining has

B.-W. Kim (ed.), *Helicobacter Infection in Clinical Practice*,
https://doi.org/10.1007/978-981-97-8397-7_6

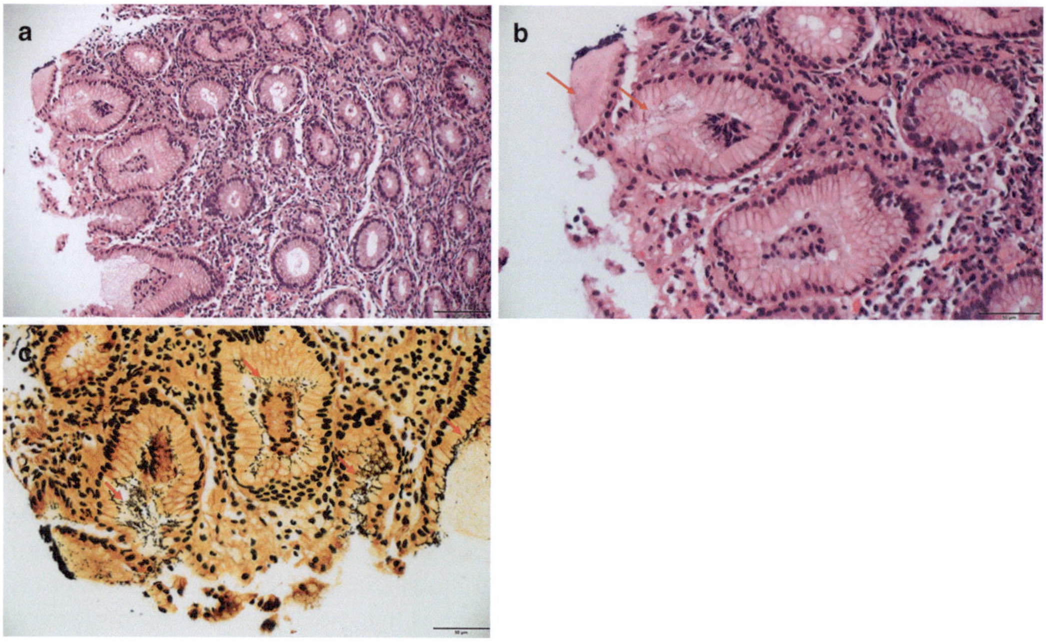

Fig. 6.1 Gastric mucosa shows chronic active inflammation (H&E, ×200) (**a**) with numerous *H. pylori* bacilli (arrow, H&E, ×400) (**b**). The *Helicobacter pylori* organisms (arrows) are present in Warthin–Starry stain (Warthin-Starry stain, ×400) (**c**)

shown enhancements in terms of time efficiency, necessitating fewer organic chemicals such as methanol and acetic acid, while still maintaining the same level of accuracy as traditional Giemsa staining. Moreover, the diagnostic efficacy of this method surpasses that of the rapid urease test [7].

In addition, immunohistochemistry (IHC) using anti-*H. pylori* antibodies has a very high sensitivity and accuracy. The IHC staining method that utilizes anti-*H. pylori* antibodies is now considered a gold standard method for detecting *H. pylori* in tissues and provides a sensitivity and specificity of approximately 100%. [8] Even when the number of *H. pylori* organisms is small, it can increase the diagnosis rate and it can also increase the diagnosis rate in cases with chronic active gastritis, atrophic gastritis, and intestinal metaplasia [9, 10]. However, it is expensive and not available in all pathology laboratories, so it can be used selectively when absolutely necessary.

These diagnostic methods using gastric biopsy tissue may have the possibility of a false-negative result depending on the biopsy location and number, as the bacteria may not be evenly distributed in the gastric mucosa [11]. Particularly, atrophic gastritis and intestinal metaplasia are not environments conducive to the survival of *H. pylori*, so the bacteria may not be found. To increase the diagnosis rate of *H. pylori*, it is recommended to collect a total of four samples, one each from the lesser and greater curvature of the antrum and body [9, 12]. However, if the number of tissue examinations needs to be reduced, it is recommended to collect tissue from areas where there is little or no atrophic gastritis and intestinal metaplasia. It is desirable to collect more than two samples each from the antrum and the body, and if a biopsy is performed in only one place, it is recommended to collect tissue from areas where there is little or no atrophic gastritis and intestinal metaplasia. Therefore, it is generally better to collect tissue from the greater curvature of the body rather than the antrum, where atrophic gastritis and intestinal metaplasia are relatively severe [13]. However, in the case of duodenal ulcers, tissue should be collected from the antrum because *H. pylori* colony formation is higher there than in the body [14].

6.2.2 Molecular Methods

The polymerase chain reaction (PCR) test is a method to enhance the detection sensitivity by amplifying the target DNA. PCR is among the most effective molecular techniques with extensive uses in the clinical field. The detection of *H. pylori* infection through PCR methods can be accomplished not only by using invasively collected gastric tissue, but also through samples of stool, saliva, dental plaques, and urine [15, 16].

The PCR test has a higher sensitivity and specificity compared to traditional methods as it can detect positivity even with a small amount of bacteria [2]. PCR-based methods are capable of identifying *H. pylori* infection in patients with peptic ulcer bleeding, gastric cancer, or gastric MALT lymphoma, having a sensitivity and specificity close to 100% [17, 18]. PCR has been demonstrated to identify active infection in a subset of healthy individuals who were confirmed as *H. pylori*-negative using traditional nonmolecular techniques [19]. Likewise, a recent study indicated that nearly half of the patients with chronic mucosal inflammation, who tested negative for *H. pylori* organisms with histological methods, were found to be infected when using PCR-based techniques [20]. Moreover, the ability to identify *H. pylori* in both spiral and coccoid forms, which is not feasible with other conventional histologic methods, is likely the most significant benefit of using the PCR method [21].

Although they are fast, accurate, and highly sensitive techniques, a notable disadvantage of these molecular methods is the potential for false-positive results as a consequence of remaining genetic components after antimicrobial treatment or the presence of other flora that generate urease or mimic genetic information [5]. In addition, they are relatively expensive and require specific techniques and equipment for testing. Molecular methods for identifying *H. pylori* encompass conventional PCR, nested PCR, multiplex PCR, fluorescence in situ hybridization (FISH), real-time PCR (RT-PCR), and digital PCR.

6.2.2.1 Conventional PCR

Conventional PCR is a process that amplifies DNA, producing millions of copies of a particular DNA segment from a very small quantity of initial substance. In conventional PCR, various genes such as *vacA*, *cagA*, *flaA*, *ureA*, *ureC*, *glmM*, 16S rRNA, 23S rRNA, and hsp60 can be targeted to confirm the presence of bacteria [22, 23]. Of these, it has been found that the amplification of the 23S rRNA gene exhibits the most effective results when it comes to detecting *H. pylori* infection [24]. Conventional PCR is usually performed with bacterial DNA isolates from stomach biopsy material. However, other specimens, such as gastric juice, saliva or stools can also be employed for the purpose of amplifying particular genes. A study that contrasted the use of gastric juice specimens with bacterial culture found that PCR using gastric juice is a superior diagnostic tool compared to the culture method, with results having high reliability and stability even in patients who had undergone multiple treatment cycles [25]. In addition, the method of identifying *H. pylori* in saliva through PCR amplification of 16S rRNA genes has proven to be reliable, with sensitivity and specificity of 80.0% and 77.7%, respectively [26]. However, several studies have indicated limited detection sensitivity in fecal samples, a limitation potentially attributable to the low quantity of target DNA and the presence of PCR-inhibiting substances within the stool samples [21, 27].

6.2.2.2 Nested PCR

Nested PCR is a method designed to minimize the nonspecific amplification of the DNA template by employing a two-step PCR process. In the initial PCR round, a broader DNA region is amplified, while the second round targets a smaller subsection of the first run product, which serves as the template [28]. Compared to conventional PCR, nested PCR shows increased sensitivity as it has the ability to amplify target DNA even at markedly lower concentrations [5, 29]. It proves effective in detecting *H. pylori* in challenging specimens, including those obtained from patients who are bleeding or on PPI medication, and can be implemented in laboratories equipped with basic facilities. Nested PCR has been shown to be at least 10 times more efficient than conventional PCR. However, it is worth noting that the likelihood of false positives due to contamination is also a risk with this technique [30].

6.2.2.3 Multiplex PCR

Multiplex PCR is the modification of conventional PCR, which allows for the concurrent amplification of several genes within a single PCR procedure. If multiplex PCR detects any one or more specific gene products such as cagA, iceA1, and vacA (subtypes: s1, s2, m1, and m2) in the specimens placed on the agarose gel, it is considered positive of *H. pylori* presence [31, 32]. When using CIM and CLO tests as the standard reference methods, multiplex PCR demonstrated a higher *H. pylori* detection rate, which underlines its superior performance for the *H. pylori* detection [33, 34]. In addition, multiplex PCR shows outstanding effectiveness in detecting *H. pylori* from fecal samples, even being capable of identifying minimal quantities of *H. pylori* genetic material [5].

6.2.2.4 Fluorescence *In Situ* Hybridization

Fluorescence *in situ* hybridization (FISH) is a highly sensitive and specific molecular cytogenetic technique that uses fluorescent-tagged oligonucleotide probes that bind to only particular parts of a nucleic acid sequence with a high degree of sequence complementarity [35]. FISH is capable of identifying coccoid forms of *H. pylori* infection. Moreover, an advantage of FISH is its ability to confirm whether the bacteria are susceptible or resistant to clarithromycin [36].

6.2.2.5 Real-Time PCR

Real-time PCR measures, in real time, the amount of PCR product molecules (amplicons) generated after each round of amplification using a fluorescent readout. In recent years, real-time PCR has become the leading tool for diagnosing *H. pylori* infections because of its short turnaround time, high sensitivity and specificity, and low risk of cross-contamination [5]. In RT-PCR, the genes most frequently targeted include fragments of 16S rRNA genes, *ureA*, *ureC*, *gyrA*, *cagA* and 23S rRNA genes, which are well-maintained in gastric tissue samples [37]. Detection of *H. pylori* using 23S rRNA RT-PCR gives completely consistent results when compared with reference methods of bacterial culture and histopathology [38]. RT-PCR is a highly sensitive and specific method to detect *H. pylori* in gastric biopsy specimens from patients with peptic ulcers [38, 39]. RT-PCR has the ability to identify *H. pylori* infections in a substantial proportion of biopsy samples that test negative histologically [40]. In an investigation using FFPE tissue samples, RT-PCR demonstrated a sensitivity of 95.6% in identifying *H. pylori* infection, a significant improvement over histological methods that displayed a sensitivity of merely 69.9% [41]. In both research and clinical environments, the FFPE tissues are the most commonly used, with highly effective and specific outcomes for the detection of *H. pylori* [42].

The use of real-time PCR and other molecular techniques is growing for the detection and typing of *H. pylori*, as well as the determination of the bacterium's susceptibility to antibiotics. Because it doesn't require viable bacteria for *H. pylori* DNA detection, the RT-PCR method is useful when a patient has already received antimicrobial treatment [5].

6.2.2.6 Digital PCR

Digital PCR (dPCR) is a specialized approach to target nucleic acids detection and quantification that estimates absolute numbers of molecules through statistical methods. dPCR is the most recent commercial evolution of PCR technology, with a sensitivity comparable to that of RT-PCR [43]. Droplet Digital PCR (ddPCR) is a method for performing digital PCR that is based on water-oil emulsion droplet technology. A sample is fractionated into thousands of droplets (10,000–100,000), and then PCR amplification of the template molecules occurs within each separate droplet [44].

The ddPCR method can identify low-density *H. pylori* infections, particularly in patients with an extremely low bacterial concentration such as peptic ulcer bleeding, gastric MALT lymphoma, or atrophic gastritis, where standard testing methods may not yield optimal results [45].

6.2.3 Molecular Methods for Antimicrobial Resistance

Although culture-based minimal inhibitory concentration (MIC) detection is considered the gold standard method, it is not commonly used because it usually requires up to 2 weeks to obtain results. Consequently, rapid and extremely precise molecular methods are gaining favor for assessing antimicrobial resistance. Recent studies investigating the practicality of molecular antibiotic susceptibility testing (AST) in relation to culture-based methods concluded that molecular techniques are reliable for diagnosing resistance to clarithromycin and levofloxacin using both bacterial and biopsy samples [46, 47]. Molecular methods can also identify heteroresistant infections; the simultaneous presence of *H. pylori* strains in the same patient that are both susceptible and resistant to the same antibiotic [48].

The most frequently utilized strategies for molecular AST involve the PCR and/or DNA hybridization based methods. Molecular AST typically relies on identifying the prevalent DNA mutations that are associated with resistance. In most instances, resistance to clarithromycin in *H. pylori* is due to three specific single point mutations (A2146C, A2146G, and A2147G) in the 23S rRNA gene, that can be precisely identified using PCR based techniques [49].

The majority of commercially available molecular AST for *H. pylori*, including the Seeplex® H. pylori-ClaR ACE Detection (Seegene, South Korea), the ClariRes real-time PCR assay (Ingentix, Austria), the *H. pylori* Taqman® real-time PCR assay (Meridian Bioscience, United States) and the RIDAGENE *H. pylori* (R-Biopharm, Germany) use real-time PCR to identify resistance to clarithromycin. The GenoType HelicoDR assay (Hain Lifescience, Germany) has the advantage of detecting mutations conferring resistance to both clarithromycin and levofloxacin [50]. In addition, Trespalacios et al. used new primers for identifying the N87I mutation in *H. pylori*'s *gyrA* gene with allele-specific PCR test to directly detect the bacterium's resistance to levofloxacin in gastric biopsy samples, which improved the sensitivity of the method for predicting levofloxacin resistance from 52% to 100% [51].

Commercially available molecular tests can detect the most prevalent DNA mutations associated with resistance, thus it is important to consider that other mutations or mechanisms might also play a part in antibiotic resistance. DNA sequencing not only allows for the detection of well-established mutations that cause antibiotic resistance, but also offers the opportunity to discover new mutations or DNA elements causing resistance, by comparing the DNA sequences of *H. pylori* strains identified as resistant or sensitive to antibiotics through culture-based methods. However, all these molecular methods only provide genetic information, without direct analysis of the amino acid sequence. Therefore, false positives could occur when a DNA mutation is detected that doesn't actually cause phenotypic resistance.

In summary, molecular biological tests are fast, convenient, highly accurate, and can confirm antibiotic susceptibility, so they are currently an actively researched method and an area of growing expectation for the future.

6.3 Conclusion

In conclusion, while histopathological examination remains an essential component in the diagnosis of *H. pylori* infection, there is considerable scope for advancement and innovation in this field. Embracing new molecular technologies and interdisciplinary approaches will be key to increasing diagnostic accuracy and patient outcomes in the future.

References

1. Braden B. Diagnosis of Helicobacter pylori infection. BMJ. 2012;344:e828.
2. Wang YK, Kuo FC, Liu CJ, Wu MC, Shih HY, Wang SS, et al. Diagnosis of Helicobacter pylori infection: current options and developments. World J Gastroenterol. 2015;21(40):11221–35.
3. Hartman DJ, Owens SR. Are routine ancillary stains required to diagnose Helicobacter infection in gastric biopsy specimens? An institutional quality assurance review. Am J Clin Pathol. 2012;137(2):255–60.
4. Lee JY, Kim N. Diagnosis of Helicobacter pylori by invasive test: histology. Ann Transl Med. 2015;3(1):10.
5. Ansari S, Yamaoka Y. Helicobacter pylori infection, its laboratory diagnosis, and antimicrobial resistance: a perspective of clinical relevance. Clin Microbiol Rev. 2022;35(3):e0025821.
6. Batts KP, Ketover S, Kakar S, Krasinskas AM, Mitchell KA, Wilcox R, et al. Appropriate use of special stains for identifying Helicobacter pylori: recommendations from the Rodger C. Haggitt Gastrointestinal Pathology Society. Am J Surg Pathol. 2013;37(11):e12–22.
7. Fan CC, Chen CH, Chou C, Kao TY, Cheng AN, Lee AY, et al. A time-saving-modified Giemsa stain is a better diagnostic method of Helicobacter pylori infection compared with the rapid urease test. J Clin Lab Anal. 2020;34(4):e23110.
8. Kocsmar E, Szirtes I, Kramer Z, Szijarto A, Bene L, Buzas GM, et al. Sensitivity of Helicobacter pylori detection by Giemsa staining is poor in comparison with immunohistochemistry and fluorescent in situ hybridization and strongly depends on inflammatory activity. Helicobacter. 2017;22(4).
9. Malfertheiner P, Megraud F, O'Morain CA, Gisbert JP, Kuipers EJ, Axon AT, et al. Management of Helicobacter pylori infection-the Maastricht V/ Florence Consensus Report. Gut. 2017;66(1):6–30.
10. Toulaymat M, Marconi S, Garb J, Otis C, Nash S. Endoscopic biopsy pathology of Helicobacter pylori gastritis. Comparison of bacterial detection by immunohistochemistry and Genta stain. Arch Pathol Lab Med. 1999;123(9):778–81.
11. el-Zimaity HM. Accurate diagnosis of Helicobacter pylori with biopsy. Gastroenterol Clin North Am. 2000;29(4):863–9.
12. Satoh K, Kimura K, Taniguchi Y, Kihira K, Takimoto T, Saifuku K, et al. Biopsy sites suitable for the diagnosis of Helicobacter pylori infection and the assessment of the extent of atrophic gastritis. Am J Gastroenterol. 1998;93(4):569–73.
13. Lee JH, Park YS, Choi KS, Kim DH, Choi KD, Song HJ, et al. Optimal biopsy site for Helicobacter pylori detection during endoscopic mucosectomy in patients with extensive gastric atrophy. Helicobacter. 2012;17(6):405–10.
14. Khulusi S, Mendall MA, Patel P, Levy J, Badve S, Northfield TC. Helicobacter pylori infection density and gastric inflammation in duodenal ulcer and non-ulcer subjects. Gut. 1995;37(3):319–24.
15. Dus I, Dobosz T, Manzin A, Loi G, Serra C, Radwan-Oczko M. Role of PCR in Helicobacter pylori diagnostics and research–new approaches for study of coccoid and spiral forms of the bacteria. Postepy Hig Med Dosw (Online). 2013;67:261–8.
16. Rimbara E, Sasatsu M, Graham DY. PCR detection of Helicobacter pylori in clinical samples. Methods Mol Biol. 2013;943:279–87.
17. He Q, Wang JP, Osato M, Lachman LB. Real-time quantitative PCR for detection of Helicobacter pylori. J Clin Microbiol. 2002;40(10):3720–8.
18. Lin HJ, Lo WC, Perng CL, Tseng GY, Li AF, Ou YH. Mucosal polymerase chain reaction for diagnosing Helicobacter pylori infection in patients with bleeding peptic ulcers. World J Gastroenterol. 2005;11(3):382–5.
19. Thorell K, Bengtsson-Palme J, Liu OH, Palacios Gonzales RV, Nookaew I, Rabeneck L, et al. In vivo analysis of the viable microbiota and Helicobacter pylori transcriptome in gastric infection and early stages of carcinogenesis. Infect Immun. 2017;85(10).
20. Kiss S, Zsikla V, Frank A, Willi N, Cathomas G. Helicobacter-negative gastritis: polymerase chain reaction for Helicobacter DNA is a valuable tool to elucidate the diagnosis. Aliment Pharmacol Ther. 2016;43(8):924–32.
21. Cardos AI, Maghiar A, Zaha DC, Pop O, Fritea L, Miere Groza F, et al. Evolution of diagnostic methods for Helicobacter pylori infections: from traditional tests to high technology, advanced sensitivity and discrimination tools. Diagnostics (Basel). 2022;12(2).
22. Calvet X, Lehours P, Lario S, Megraud F. Diagnosis of Helicobacter pylori infection. Helicobacter. 2010;15(Suppl 1):7–13.
23. Ricci C, Holton J, Vaira D. Diagnosis of Helicobacter pylori: invasive and non-invasive tests. Best Pract Res Clin Gastroenterol. 2007;21(2):299–313.
24. Khadangi F, Yassi M, Kerachian MA. Review: Diagnostic accuracy of PCR-based detection tests for Helicobacter Pylori in stool samples. Helicobacter. 2017;22(6).
25. Hsieh MS, Liu CJ, Hsu WH, Li CJ, Tsai PY, Hu HM, et al. Gastric juice-based PCR assay: an alternative testing method to aid in the management of previously treated Helicobacter pylori infection. Helicobacter. 2019;24(2):e12568.
26. Sekhar Goud EVS, Kannan R, Rao UK, Joshua E, Tavaraja R, Jain Y. Identification of Helicobacter pylori in saliva of patients with and without gastritis by polymerase chain reaction. J Pharm Bioallied Sci. 2019;11(Suppl 3):S523–S9.
27. Lottspeich C, Schwarzer A, Panthel K, Koletzko S, Russmann H. Evaluation of the novel Helicobacter pylori ClariRes real-time PCR assay for detection and clarithromycin susceptibility testing of H. pylori

in stool specimens from symptomatic children. J Clin Microbiol. 2007;45(6):1718–22.
28. Green MR, Sambrook J. Nested polymerase chain reaction (PCR). Cold Spring Harb Protoc. 2019;2019(2).
29. Singh V, Mishra S, Rao GR, Jain AK, Dixit VK, Gulati AK, et al. Evaluation of nested PCR in detection of Helicobacter pylori targeting a highly conserved gene: HSP60. Helicobacter. 2008;13(1):30–4.
30. Seligova B, Lukac L, Babelova M, Vavrova S, Sulo P. Diagnostic reliability of nested PCR depends on the primer design and threshold abundance of Helicobacter pylori in biopsy, stool, and saliva samples. Helicobacter. 2020;25(2):e12680.
31. Salari MH, Shirazi MH, Hadaiti MA, Daryani NA. Frequency of Helicobacter pylori vacA genotypes in Iranian patients with gastric and duodenal ulcer. J Infect Public Health. 2009;2(4):204–8.
32. Oktem-Okullu S, Tiftikci A, Saruc M, Cicek B, Vardareli E, Tozun N, et al. Multiplex-PCR-based screening and computational modeling of virulence factors and T-cell mediated immunity in Helicobacter pylori infections for accurate clinical diagnosis. PLoS One. 2015;10(8):e0136212.
33. Fadilah N, Hanafiah A, Razlan H, Wong ZQ, Mohamed Rose I, Rahman MM. Multiplex PCR for detection of Helicobacter pylori infection in gastric biopsies with lower inflammatory score. Br J Biomed Sci. 2016;73(4):180–7.
34. Trung TT, Minh TA, Anh NT. Value of CIM, CLO test and multiplex PCR for the diagnosis of Helicobacter pylori infection status in patients with gastritis and gastric ulcer. Asian Pac J Cancer Prev. 2019;20(11):3497–503.
35. Hu L, Ru K, Zhang L, Huang Y, Zhu X, Liu H, et al. Fluorescence in situ hybridization (FISH): an increasingly demanded tool for biomarker research and personalized medicine. Biomark Res. 2014;2(1):3.
36. Tajbakhsh S, Samarbaf-Zadeh AR, Moosavian M. Comparison of fluorescent in situ hybridization and histological method for the diagnosis of Helicobacter pylori in gastric biopsy samples. Med Sci Monit. 2008;14(9):BR183–7.
37. Schabereiter-Gurtner C, Hirschl AM, Dragosics B, Hufnagl P, Puz S, Kovach Z, et al. Novel real-time PCR assay for detection of Helicobacter pylori infection and simultaneous clarithromycin susceptibility testing of stool and biopsy specimens. J Clin Microbiol. 2004;42(10):4512–8.
38. Ramirez-Lazaro MJ, Lario S, Casalots A, Sanfeliu E, Boix L, Garcia-Iglesias P, et al. Real-time PCR improves Helicobacter pylori detection in patients with peptic ulcer bleeding. PLoS One. 2011;6(5):e20009.
39. Lo CC, Lai KH, Peng NJ, Lo GH, Tseng HH, Lin CK, et al. Polymerase chain reaction: a sensitive method for detecting Helicobacter pylori infection in bleeding peptic ulcers. World J Gastroenterol. 2005;11(25):3909–14.
40. Benejat L, Ducournau A, Lehours P, Megraud F. Real-time PCR for Helicobacter pylori diagnosis. The best tools available. Helicobacter. 2018;23(5):e12512.
41. Liu Q, Qi D, Kang J, Jin Y, Liu W, Gao W, et al. Efficacy of real-time PCR-based detection of Helicobacter pylori infection and genotypic resistance-guided quadruple therapy as the first-line treatment for functional dyspepsia with Helicobacter pylori infection. Eur J Gastroenterol Hepatol. 2015;27(3):221–5.
42. Ciesielska U, Jagoda E, Marciniak Z. Value of PCR technique in detection of Helicobacter pylori in paraffin-embedded material. Folia Histochem Cytobiol. 2002;40(2):129–30.
43. Quan PL, Sauzade M, Brouzes E. dPCR: a technology review. Sensors (Basel). 2018;18(4).
44. Taylor SC, Laperriere G, Germain H. Droplet Digital PCR versus qPCR for gene expression analysis with low abundant targets: from variable nonsense to publication quality data. Sci Rep. 2017;7(1):2409.
45. Ramirez-Lazaro MJ, Lario S, Quilez ME, Montserrat A, Bella MR, Junquera F, et al. Droplet digital PCR detects low-density infection in a significant proportion of Helicobacter pylori-negative gastric biopsies of dyspeptic patients. Clin Transl Gastroenterol. 2020;11(6):e00184.
46. Zhou J, Shen Y, Song X, Zhou L, Tang H, Li H. Evaluation of a molecular mosprie assay for detection of Helicobacter pylori and resistance to clarithromycin and levofloxacin. J Infect Dis. 2022;226(Suppl 5):S503–S9.
47. Wang YH, Li Z, Wang L, Zhu-Ge LY, Zhao RL, Wu S, et al. A systematic review and meta-analysis of genotypic methods for detecting antibiotic resistance in Helicobacter pylori. Helicobacter. 2018;23(2):e12467.
48. Kao CY, Lee AY, Huang AH, Song PY, Yang YJ, Sheu SM, et al. Heteroresistance of Helicobacter pylori from the same patient prior to antibiotic treatment. Infect Genet Evol. 2014;23:196–202.
49. Redondo JJ, Keller PM, Zbinden R, Wagner K. A novel RT-PCR for the detection of Helicobacter pylori and identification of clarithromycin resistance mediated by mutations in the 23S rRNA gene. Diagn Microbiol Infect Dis. 2018;90(1):1–6.
50. Cambau E, Allerheiligen V, Coulon C, Corbel C, Lascols C, Deforges L, et al. Evaluation of a new test, genotype HelicoDR, for molecular detection of antibiotic resistance in Helicobacter pylori. J Clin Microbiol. 2009;47(11):3600–7.
51. Trespalacios AA, Rimbara E, Otero W, Reddy R, Graham DY. Improved allele-specific PCR assays for detection of clarithromycin and fluoroquinolone resistant of Helicobacter pylori in gastric biopsies: identification of N87I mutation in GyrA. Diagn Microbiol Infect Dis. 2015;81(4):251–5.

7 When to Perform Noninvasive Test for *Helicobacter pylori*?

Jun-Won Chung

Highlights

What Is Known
• Noninvasive testing for *H. pylori* refers to methods that don't involve invasive procedures or tools. Although various noninvasive tests for *H. pylori* are commonly used in clinical practice, there is not yet a universally accepted gold standard method for diagnosing *H. pylori* infection.
Future Research Questions
• Currently, invasive methods for gastric tissue are commonly employed, but research into noninvasive approaches for detecting and confirming resistance is actively underway.

7.1 Introduction

Helicobacter pylori (*H. pylori*) is a motile, spiral- or curved rod-shaped Gram-negative bacterium with multiple flagella, residing in the gastric mucous-coated lining and the gastric pits of the epithelial tissue of the stomach. A noninvasive *H. pylori* test refers to techniques that do not require any endoscopic tools. While several noninvasive *H. pylori* tests are in routine clinical use, there is currently no single method that can be unequivocally regarded as the gold standard for diagnosis of *H. pylori* infection.

Clinical conditions, availability, and costs should be taken into account when selecting the most suitable test. The primary clinical scenario for noninvasive tests is patients who cannot tolerate endoscopic examination. Noninvasive tests, such as urea breath test (UBT) and serologic test, can provide comprehensive assessment of the entire stomach. On the other hand, invasive tests may reflect the specific focal conditions within the stomach [1, 2].

7.2 Main Body

7.2.1 Urea Breath Test

The UBT is among the most commonly used non-invasive tests for *H. pylori*. Numerous pieces of evidences support the utility of UBT in assessing the eradication status. In fact, most guidelines recommend UBT as the preferred method to confirm the eradication of infection and to determine the infection status in patients experiencing recurrent symptoms after eradication.

The UBT is commonly utilized to verify the eradication after treatment of *H. pylori* especially in cases like duodenal ulcer patients who do not require biopsies for the ulcer. However, in specific situations such as gastric ulcer or gastric mucosa-associated lymphoid tissue (MALT)

J.-W. Chung (✉)
Department of Internal Medicine, Gil Medical Center, Gacheon University, Incheon, Korea

B.-W. Kim (ed.), *Helicobacter Infection in Clinical Practice*,
https://doi.org/10.1007/978-981-97-8397-7_7

lymphoma, follow-up necessitates upper gastrointestinal endoscopy, and biopsy-based tests should be conducted to confirm *H. pylori* eradication. In other situations, a noninvasive test is employed.

The sensitivity and specificity of UBT are comparable to invasive tests for diagnosing *H. pylori* infection; however, it does not reveal associated complications such as gastroduodenal ulcers/erosions, dysplasia, and neoplasms. The ^{13}C-UBT boasts high accuracy, is easy to conduct, and stands as the best optimal test for diagnosing *H. pylori* infection. Nevertheless, it has demonstrated varying accuracy levels in the pediatric age group, particularly in young children [3].

However, there is a drawback to the UBT. A common issue arises in the form of results falling within a gray area. The widely used UBT indicates *H. pylori* infection with an increase in the ^{13}C:^{12}C ratio (delta$^{13}CO_2$) of 2.5% after urea intake. However, the 2.5% cutoff value resulted in an unexpected rate of false-positive results. To address this, a cutoff value between 2.5% and 10% was designated as the gray zone, and ultimately, a cutoff value of 8.5% was chosen to maintain high sensitivity and acceptable specificity [4].

To avoid false-negative test results, a study suggested that the incorporation of an acidic test meal could overcome the issue. Anti-acidic drug such as H_2 blockers (H2RAs), proton pump inhibitors (PPIs), and potassium competitive acid blockers (P-CABs) may also contribute to false-negative results; therefore, it is crucial to discontinue their use 2 weeks prior to testing. However, stopping these medications can pose challenges for some patients with acid-related disorder.

7.2.2 Serologic Tests

The serologic test is one of the most commonly utilized non-invasive methods for diagnosing *H. pylori* infection. It is frequently employed in epidemiologic studies due to its cost-effectiveness and noninvasiveness. Detecting specific antibodies developed after exposure to various *H. pylori* antigens proves to be a valuable approach in clinical practices.

In situations where patients are undergoing treatment with PPIs and discontinuation is not feasible, a serologic test can provide highly beneficial. Another clinical situation in which serologic test is extensively employed is in cases of stomach bleeding, atrophic gastritis, and metaplastic gastritis, especially when other diagnostic methods yield false-negative results.

The drawback of serologic test is that it may not be ideal for confirming the eradication status. Antibodies against *H. pylori* can persist at elevated levels for an extended period even after successful eradication.

7.2.3 Stool Antigen Test

Confirming the presence of *H. pylori* antigen in stool enables the development of stool antigen test (SAT). The evolution of this immunologic assay, transitioning from polyclonal to monoclonal antibodies, has significantly improved sensitivity and specificity.

The primary application is in pediatric patients, particularly as conducting the UBT is challenging for young children. Another advantage of SAT is its utility in confirming the eradication of the infection through a straightforward laboratory examination. Targeting pediatric populations, SAT employing monoclonal antibodies offers improved feasibility due to its cost-effectiveness and ease to handling [5].

SAT is a precise method, but its accuracy is affected by various clinical factors, including the use of potent anti-acid drugs (PPIs and P-CABs), upper gastrointestinal bleeding, the administration of antibiotics, and the consistency of the stool. Watery stool may lead to false-negative results due to antigen dilution.

7.2.4 Stool PCR

Polymerase chain reaction (PCR) is a reliable technology and is widely used in both clinical practice and medical research. Given the escalat-

Table 7.1 Noninvasive tests for *H. pylori*

	Advantages	Disadvantages
Urea breath test	Useful after *H. pylori* eradication therapy	Gray zone False negative if PPI and antibiotics is used
Serology test	Widely available Nonexpensive Useful for epidemiologic study	Not used after eradication therapy
Stool antigen test	Useful for pediatric patients	Stool collection is annoying for patients
Stool PCR	Very sensitive and specific Know antibiotics resistance pattern	Expensive Stool collection

PPI proton-pump inhibitor, *PCR* polymerase chain reaction

ing prevalence of antibiotics resistance in the real world, understanding the antibiotic resistance pattern, particularly for clarithromycin, is crucial before initiating eradication therapy. In one study, it was demonstrated that a PCR-based assay can simultaneously identify the presence of *H. pylori* DNA and mutations associated with clarithromycin resistance [6]. This approach is anticipated to extend to other antibiotics such as fluoroquinolones, amoxicillin, and rifabutin, serving as a guide for antibiotic regimens. However, the implementation of these tests remains costly and requires specialized equipment.

7.3 Conclusion

There is no unique method serving as the gold standard for diagnosing *H. pylori* infection. Consequently, it is essential to choose diagnostic methods based on the diverse clinical circumstances of individual patients. The advantages and disadvantages of each method are summarized in Table 7.1.

References

1. Gatta L, Ricci C, Tampieri A, et al. Non-invasive techniques for the diagnosis of *Helicobacter pylori* infection. Clin Microbiol Infect. 2003;9(6):489–96.
2. Stefano K, Aloe R, Bonaguri C, et al. Non-invasive tests for the diagnosis of *helicobacter pylori*: state of the art. Acta Biomed. 2018;89(8):58–64.
3. Imrie C, Rowland M, Bourke B, et al. Limitations to carbon13-labeled urea breath testing for H. pylori in infants. J Pediatr. 2001;139:734–7.
4. Calvet X, et al. Accuracy of diagnostic tests for Helicobacter pylori: a reappraisal. Clin Infect Dis. 2009;15(48):1258–91.
5. Vaira D, Malfertheiner P, Megraud F, et al. Diagnosis of Helicobacter pylori infection with a new non-invasive antigen-based assay. Lancet. 1999;354(9172):30–3.
6. Beckman E, Saracino I, Fiorini G, et al. A novel stool PCR test for Helicobacter pylori may predict clarithromycin resistance and eradication of infection at a high rate. J Clin Microbiol. 2017;55(8):2400–5.

8 Serologic Diagnosis

Tae Ho Kim

Highlights

What Is Known
• Serologic tests are valuable diagnostic tools for identifying *H. pylori* infection by detecting specific antibodies in a patient's serum, offering a noninvasive alternative to endoscopy. They are cost-effective and accessible, suitable for large-scale screening and epidemiological studies, especially in high-prevalence areas. Despite their advantages, serologic tests have limitations, including potential false results and an inability to distinguish between current and past infections. Sensitivity and specificity vary among different populations and geographic regions. Understanding the timing of testing in relation to infection or treatment is crucial for accurate results. Healthcare providers should manage patient expectations and consider the clinical context when interpreting serologic test results. Further research and adherence to evolving guidelines are necessary for optimizing the use of serologic tests in *H. pylori* diagnosis and management.
Future Research Questions
• Need for research aimed at establishing a novel serologic test capable of distinguishing current and past infections, with demonstrating satisfactory sensitivity and specificity, particularly in diverse patient populations

T. H. Kim (✉)
Department of Internal Medicine, Subdivision of Gastroenterology, Bucheon St. Mary's Hospital, The Catholic University of Korea, Seoul, Korea
e-mail: kimtaeho@songeui.ac.kr

8.1 Introduction

Serologic tests serve as diagnostic methods to identify specific antibodies or antigens in a patient's serum, typically indicating past or present infection or exposure to a particular pathogen, such as *Helicobacter pylori* (*H. pylori*). In diagnosing *H. pylori* infection, serologic tests target the detection of antibodies produced by the patient's immune system in response to *H. pylori* exposure [1, 2]. One notable advantage of serologic tests is their noninvasiveness. Unlike invasive methods like endoscopy, which requires the insertion of a tube into the stomach, serologic tests only necessitate a blood sample. This makes them particularly attractive to patients wary of more invasive procedures [3].

Serologic tests are generally cost-effective compared to invasive diagnostic methods like endoscopy or the urea breath test (UBT) [4]. The relatively low cost of serologic tests renders it an accessible option for both patients and healthcare systems. Additionally, serologic tests can be conducted in standard clinical laboratories without the need for expensive equipment or highly specialized personnel, further enhancing their cost-effectiveness [5, 6]. These characteristics make serologic tests well-suited for large-scale population screening and epidemiological studies, especially in areas with a high prevalence of *H. pylori* infection or situations requiring public health interventions [7, 8].

B.-W. Kim (ed.), *Helicobacter Infection in Clinical Practice*,
https://doi.org/10.1007/978-981-97-8397-7_8

Serologic tests are particularly useful for diagnosing *H. pylori* infections in pediatric and geriatric populations, where invasive procedures like endoscopy may be less suitable. However, it is worth noting that serologic tests may be less reliable in pediatric populations due to different immune responses, and children may not produce detectable antibody levels as quickly as adults [2, 9].

While serologic tests offer advantages, it's essential to recognize their limitations, including the inability to distinguish between current and past infections and the potential for false results. False-positive results are a concern, as individuals without an active *H. pylori* infection may still test positive for antibodies. This can result from cross-reactivity with other antigens or from successfully treated previous *H. pylori* infections. False-negative results are also possible, especially in the early stages of infection when the patient has not yet produced a significant amount of antibodies [3].

Patients with compromised immune systems, such as those with HIV/AIDS or taking immunosuppressive medications, may not generate detectable antibody levels, leading to false negatives. Medications like proton-pump inhibitors (PPIs) and antibiotics can influence antibody production, affecting the accuracy of serologic tests. False negatives can result in delayed or missed diagnoses.

The performance of serologic tests may vary significantly among different populations and geographic regions due to differences in the prevalence of *H. pylori* and variations in individuals' immune responses from diverse backgrounds. For example, in regions with a high prevalence of *H. pylori*, serologic tests may exhibit better performance due to the increased likelihood of true-positive results. Conversely, in areas with low prevalence, false positives may become more prevalent, diminishing the overall accuracy of the test. Clinicians should take these factors and the clinical context into account when determining the appropriate diagnostic method for a specific patient [4].

The timing of serologic testing in relation to *H. pylori* exposure or treatment is crucial. If testing takes place immediately after infection or treatment, antibody levels may not be detectable yet or may still be present, leading to false-negative or false-positive results, respectively [10].

Healthcare providers must inform patients about the limitations of serologic tests, as some patients may anticipate a definitive diagnosis from a single blood test. Failure to manage patient expectations can result in confusion and dissatisfaction.

8.2 Main Body

8.2.1 Serologic Test Methods

There are several methods for serologic test. It is crucial to recognize that the sensitivity and specificity can vary based on the specific serologic test employed, the tested population, and the prevalence of *H. pylori* in that particular population [11]. Comparison of each methods are summarized in Table 8.1.

8.2.1.1 Enzyme-Linked Immunosorbent Assay [8, 10, 12–14]

Methodology: ELISA is an extensively employed serologic test wherein *H. pylori*-specific antibodies in a patient's serum bind to *H. pylori* antigens coated on a microplate. The ensuing reaction is identified through an enzyme-conjugated secondary antibody.

Sensitivity: The sensitivity of ELISA typically ranges from 80% to 95%, indicating its ability to accurately identify *H. pylori* infection in 80–95% of individuals.

Specificity: The specificity of ELISA also typically ranges from 80% to 95%, indicating its accuracy in correctly identifying individuals without *H. pylori* infection at a rate of 80–95%.

8.2.1.2 Western Blot [3, 9, 13]

Methodology: Western blot is a confirmatory test employing *H. pylori* lysates, which detects multiple antigenic bands specific to *H. pylori* antibodies in a patient's serum.

Table 8.1 Different serologic test methods for detecting *Helicobacter pylori* infection

Serologic test method	Methodology	Sensitivity	Specificity
Enzyme-linked immunosorbent assay (ELISA) [8, 10, 12–14]	ELISA is an extensively employed serologic test wherein *H. pylori*-specific antibodies in a patient's serum bind to *H. pylori* antigens coated on a microplate	80–95%	80–95%
Western blot [3, 9, 13]	Western blot is a confirmatory test employing *H. pylori* lysates, which detects multiple antigenic bands specific to *H. pylori* antibodies in a patient's serum	>95%	>95%
Immunochromatographic tests [9, 10, 15]	Immunochromatographic tests are rapid point-of-care tests that employ a test strip containing *H. pylori* antigens	70–90%	70–90%
Multiplex Serology [13, 14, 16, 17]	Multiplex serology allows for the simultaneous detection of antibodies to multiple *H. pylori* antigens	>90%	>90%

Sensitivity: Western blot tests typically exhibit high sensitivity, often exceeding 95%, indicating their capacity to accurately identify over 95% of individuals with *H. pylori* infection.

Specificity: The specificity of Western blot tests is generally high, typically exceeding 95%, indicating their accuracy in correctly identifying over 95% of individuals without *H. pylori* infection.

8.2.1.3 Immunochromatographic Tests [9, 10, 15]

Methodology: Immunochromatographic tests are rapid point-of-care tests that employ a test strip containing *H. pylori* antigens. A patient's serum is applied to the strip, and the appearance of a visible line indicates the presence of antibodies.

Sensitivity: Sensitivity for immunochromatographic tests can vary, often ranging from 70% to 90%.

Specificity: Their specificity typically ranges from 70% to 90%.

8.2.1.4 Multiplex Serology [13, 14, 16, 17]

Methodology: Multiplex serology allows for the simultaneous detection of antibodies to multiple *H. pylori* antigens. This method can provide a more comprehensive diagnostic profile.

Sensitivity: The sensitivity of multiplex serology is generally high, exceeding 90%.

Specificity: Multiplex serology also offers high specificity, generally exceeding 90%.

It is important to note that the sensitivity and specificity values provided here are approximate ranges and may vary based on the specific test kit, laboratory, and patient population. Additionally, sensitivity and specificity can be influenced by factors such as the timing of the test in relation to *H. pylori* exposure and treatment.

When selecting a serologic test for the diagnosis of *H. pylori*, healthcare providers should consider the context, prevalence of *H. pylori* in the patient population, and the specific clinical requirements. Furthermore, these tests are most valuable when integrated into a comprehensive diagnostic approach, which may include clinical assessment and, in some cases, confirmatory tests such as endoscopy or urea breath tests.

8.2.2 Clinical Utility of Serologic Tests

Serologic tests for *H. pylori* have several practical applications in clinical practice. They are commonly employed for the initial diagnosis of *H. pylori* infection. They provide a noninvasive and cost-effective means of identifying individuals with an active *H. pylori* infection [10]. When a patient presents with symptoms such as dyspepsia, epigastric pain, or unexplained weight loss, serologic testing can help determining whether *H. pylori* infection is a likely cause. Additionally, serologic tests prove valuable for large-scale population screening programs, especially in regions with a high prevalence of *H.*

pylori infection. This screening can help identify individuals at risk for *H. pylori*-associated gastrointestinal diseases, including peptic ulcers and gastric cancer.

Screening programs may be conducted in schools, workplaces, or community healthcare settings to reach a broader population. After *H. pylori* eradication, which typically involves antibiotics and acid-suppressing medications, the urea breath test is commonly employed to access treatment effectiveness. However, serologic tests can also serve as a monitoring tool for treatment effectiveness. A decrease in antibody levels in follow-up serologic tests can indicate the successful eradication of the infection, reflecting a positive treatment outcome.

Serologic tests are applicable to asymptomatic individuals at risk for *H. pylori* infection, such as those with a family history of gastric cancer or those residing in high-prevalence regions. Identifying asymptomatic carriers is essential for early intervention and the prevention of related health issues. Researchers and public health agencies utilize serologic tests for conducting epidemiological studies on *H. pylori* infection. These studies contribute to understanding the distribution and prevalence of *H. pylori* in specific regions and populations. Epidemiological data derived from these studies can inform public health strategies and interventions [8].

8.3 Conclusion

Serologic tests are generally recommended as the initial screening tool for *H. pylori* infection in patients with symptoms such as dyspepsia or unexplained abdominal pain. They are particularly utilized in low-prevalence areas or when immediate endoscopy is not indicated. Clinical guidelines emphasize the importance of considering the clinical context when employing serologic tests [18, 19]. The presence of alarm symptoms or the patient's risk factors may warrant further investigation, such as endoscopy, even if serologic test results are negative.

Serologic tests are considered suitable for use in pediatric and geriatric populations due to their noninvasive nature. However, caution is advised in interpreting results, especially in children, as they may exhibit a different immune response. Clinical guidelines emphasize understanding the sensitivity and specificity of the specific serologic test being used. Test performance may vary by the test kit and the population in which it is applied. Serologic tests should be validated for the specific population they are intended for, taking into account the local prevalence of *H. pylori* infection.

Healthcare providers are encouraged to educate patients about the limitations of serologic tests, especially regarding their inability to distinguish current from past infections. Patients should be informed that a positive serologic result does not necessarily indicate an active infection. For accurate results, patients taking medications that can affect serologic test results, such as proton-pump inhibitors (PPIs), should discontinue them before testing.

It is important to reiterate that guidelines and recommendations may change over time as new research and diagnostic techniques become available. Therefore, healthcare providers should refer to the most current guidelines in their specific region or healthcare setting for the latest information on the use of serologic tests in *H. pylori* diagnosis and management.

References

1. Graham DY, Dang BN, El-Serag HB. Helicobacter pylori infection. N Engl J Med. 2019;381:587–8.
2. Crowe SE. Helicobacter pylori infection. N Engl J Med. 2019;380:1158–65.
3. Herbrink P, van Doorn LJ. Serological methods for diagnosis of Helicobacter pylori infection and monitoring of eradication therapy. Eur J Clin Microbiol Infect Dis. 2000;19:164–73.
4. Guevara B, Cogdill AG. Helicobacter pylori: a review of current diagnostic and management strategies. Dig Dis Sci. 2020;65:1917–31.

5. Mumtaz K, Abid S, Yakoob J, et al. An office-based serological test for detection of current Helicobacter pylori infection: is it useful? Eur J Gastroenterol Hepatol. 2006;18:85–8.
6. Pak K, Junga Z, Mertz A, et al. The patterns and associated cost of serologic testing for Helicobacter pylori in the U.S. Military Health System. Mil Med. 2020;185:e1417–9.
7. El-Serag HB, Kao JY, Kanwal F, et al. Houston Consensus Conference on testing for Helicobacter pylori infection in the United States. Clin Gastroenterol Hepatol. 2018;16:992–1002.e6.
8. Mentis A, Lehours P, Mégraud F. Epidemiology and diagnosis of Helicobacter pylori infection. Helicobacter. 2015;20(Suppl 1):1–7.
9. Treepongkaruna S, Nopchinda S, Taweewongsounton A, et al. A rapid serologic test and immunoblotting for the detection of Helicobacter pylori infection in children. J Trop Pediatr. 2006;52:267–71.
10. Lopes AI, Vale FF, Oleastro M. Helicobacter pylori infection – recent developments in diagnosis. World J Gastroenterol. 2014;20:9299–313.
11. Huh CW, Kim BW. Diagnosis of Helicobacter pylori infection. Korean J Gastroenterol. 2018;72:229–36.
12. Cohen H, Rose S, Lewin DN, et al. Accuracy of four commercially available serologic tests, including two office-based tests and a commercially available 13C urea breath test, for diagnosis of Helicobacter pylori. Helicobacter. 1999;4:49–53.
13. Patel SK, Pratap CB, Jain AK, et al. Diagnosis of Helicobacter pylori: what should be the gold standard? World J Gastroenterol. 2014;20:12847–59.
14. Wang YK, Kuo FC, Liu CJ, et al. Diagnosis of Helicobacter pylori infection: current options and developments. World J Gastroenterol. 2015;21:11221–35.
15. Godbole G, Mégraud F, Bessède E. Review: Diagnosis of Helicobacter pylori infection. Helicobacter. 2020;25(Suppl 1):e12735.
16. Michel A, Waterboer T, Kist M, et al. Helicobacter pylori multiplex serology. Helicobacter. 2009;14:525–35.
17. Butt J, Blot WJ, Shrubsole MJ, et al. Performance of multiplex serology in discriminating active vs past Helicobacter pylori infection in a primarily African American population in the southeastern United States. Helicobacter. 2020;25:e12671.
18. Katelaris P, Hunt R, Bazzoli F, et al. Helicobacter pylori World Gastroenterology Organization Global Guideline. J Clin Gastroenterol. 2023;57:111–26.
19. Malfertheiner P, Megraud F, O'Morain C, et al. Current concepts in the management of Helicobacter pylori infection: the Maastricht III Consensus Report. Gut. 2007;56:772–81.

Urea Breath Test

9

Cheol Min Shin

Highlights

What Is Known
• The urea breath test (UBT) evaluates *Helicobacter pylori* urease activity, *via* detecting labeled carbon dioxide in the breath after the ingestion of radioactive isotope (C^3 or C^4) labeled urea. • UBT is a noninvasive, simple, and safe test with excellent accuracy for the diagnosis of *H. pylori* infection as well as for the confirmation of successful *H. pylori* eradication.
Future Research Questions
• The test protocol of the UBT is not completely standardized, especially urea dose, whether or not fasting before test is necessary, whether a test meal should be used, and type of the test meal. • Diagnostic accuracy of the UBT can be variable in various clinical settings. More research is required on the optimal protocol for conducting the UBT and interpreting the test results in various clinical situations, that is, patients with partial gastrectomy or pediatric patients.

9.1 Introduction

To overcome this limitation of serology and invasive methods to diagnose *H. pylori* infection, the urea breath test (UBT) using ^{13}C or ^{14}C-labeled urea has been developed [1–4]. The ^{13}C-UBT was first introduced for the detection of *Helicobacter pylori* infection in 1987 [5]. This test has been widely used because it showed a sensitivity and specificity over 90% for detecting *H. pylori* infection, easy to repeat, highly accurate, and to be a simple and convenient method for the patients [2, 6]. It has become popular in clinical practice both clinical conditions where endoscopy is not necessary and especially to check the success of eradication regimens.

9.2 Main Body

9.2.1 The Principle of the Urea Breath Test

Using the specific biochemical properties of *H. pylori*, $^{13/14}C$-UBT can be applied to diagnose the presence of the organism in the gastric or duodenal mucosa. *H. pylori* produces urease that hydrolyzes urea into ammonium and bicarbonate. Urease activity, then, seems to be essential to the survival of *H. pylori* in the stomach. The production of ammonium and bicarbonate leads to local buffering of gastric acid (pH 1.5) and gives *H. pylori* a unique microbiological ability to survive in the hostile intra-gastric environment [7]. Bicarbonate produced in the gastric mucosa enters the bloodstream, is transported to the lungs, and then expelled in the form of carbon dioxide when exhaled. In the $^{13/14}C$-UBT, a urea

C. M. Shin (✉)
Department of Internal Medicine, Seoul National University Bundang Hospital, Seongnam, Korea

B.-W. Kim (ed.), *Helicobacter Infection in Clinical Practice*,
https://doi.org/10.1007/978-981-97-8397-7_9

capsule tagged with radioactive carbon (^{13}C or ^{14}C) is administered orally to a patient, then the patient's exhaled carbon dioxide is collected in airtight balloons. Detection of markedly elevated proportions of exhaled $^{13/14}CO_2$ indicates marked bacterial urease activity in the patient's stomach, which is a biologically plausible indirect measure of *H. pylori* presence [8].

The ^{13}C-labeled UBT is generally recommended by the guidelines and several expert consensus reports over ^{14}C-labeled urea since ^{13}C is a stable, nonradioactive carbon isotope with a natural abundance of approximately 1% among all carbon isotopes [9]. However, the measurement of ^{13}C-labeled carbon dioxide by conventional methods is more expensive and requires skilled personnel. Therefore, other alternatives have been developed that use inexpensive methods such as infrared spectroscope and laser-assisted ratio analyzer to measure the ^{13}C-labeled carbon dioxide activity.

9.2.2 How to Perform the Urea Breath Test

In general, fasting for at least 4 h before ^{13}C-UBT is recommended. Some studies reported that fasting did not cause differences in UBT test results [10, 11], but other studies showed a clear decrease in delta-over-baseline (DOB) 30 min after a meal, emphasizing that it can be an important cause of false-negative results [12]. Caution is required in patients taking antibiotics or proton-pump inhibitors (PPIs) or in patients with upper gastrointestinal bleeding, owing to the possibility of false-negative results. In general, it is recommended to discontinue PPIs for at least 2 weeks and antibiotics for 4 weeks or more before the UBT.

In general, breath samples during the UBT are collected twice: before and after urea ingestion. In the original protocol [5], breath samples were collected every 10 min for 3 h. However, the test has been simplified using the protocol by measuring $^{13}CO_2$ at 20–30 min after urea ingestion. It has been shown that post breath sampling too early can lead to false-positive results due to the urease activity of oral bacteria [5, 13]. On the other hand, sampling too late may produce false-negative results. Accuracy with the post breath sampling at 20–30 min was reportedly very high, with both sensitivity and specificity of ~97%. When UBT is performed following the most widely used protocol with citric acid and 75 mg of urea solution, excellent accuracy results (sensitivity of 99% and specificity of 98%) are obtained if post breath samples are collected in 10–15 min after urea ingestion [2].

9.2.3 General Considerations: Urea Substrate, Measuring Equipment, and Test Meal

With regard to urea doses, the first UBT were performed with 350 mg of urea [5]. However, UBT can be performed with relatively low doses of urea, that is, 100 mg, 75 mg, or even 50 mg, which shows a good performance of a sensitivity and specificity of ~97% [14, 15]. The most widely used protocols (100 mg without citric acid or 75 mg with citric acid) allow excellent accuracy when collecting breath samples in 10–15 min after urea administration [16]. Urease-producing bacteria in the oral cavity may theoretically cause false-positive UBT results. To overcome this problem, urea capsule or tablet can be used for the test [15, 17–20].

Urea can be labeled with either ^{14}C or ^{13}C. The main difference between the two isotopes is that ^{14}C is radioactive, whereas ^{13}C is not [21]. ^{14}C-urea is cheap and rapid, so the ingestion of ^{14}C-urea capsule allows an accurate diagnostic performance using a single 10-min breath sample [14, 22]. It does not require any test meal. While the ^{14}C-UBT test can now be performed with 1 μCi dose, equivalent to the natural background radiation exposure in a day [9, 21], the use of ^{14}C-UBT is limited due to the requirement of a nuclear medicine department or centers licensed for storage and disposal of radioactive substrates. In contrast, ^{13}C is a nonradioactive isotope that can be used safely for repeated testing. It is suitable for detecting *H. pylori* infection in children [21]. However, the major limitations of ^{13}C-urea are the higher cost compared with ^{14}C-urea, and the need for expensive mass spectrometry [21].

After hydrolysis of ^{13}C-labeled urea by *H. pylori* urease in the stomach, $^{13}CO_2$ is transported to the lungs *via* bloodstream. ^{13}C in exhaled air can be detected by specific measuring equipment. ^{13}C is measured as the $^{13}CO_2$:$^{12}CO_2$ isotope ratio and is expressed as DOB per mil (‰) with respect to the international reference standard represented by the Pee Dee Belemnite limestone [21]. ^{13}C measuring equipment includes isotope ratio mass spectrometer (IRMS), nondispersive isotope selective infrared spectroscope (NDIRS), and laser-assisted ratio analyzer (LARA) [23]. The advantage of IRMS is that it processes the highest number of breath samples and works in an automated manner. However, it is inexpensive and requires the longest analysis time. NDIRS is cheaper, and it does not require helium as the carrier gas, but it can analyze only a small number of samples. The LARA system has the fastest analysis time. In general, IRMS and LARA are more suitable for gastroenterological centers requiring large quantity, automated analysis, whereas NDIRS is more suitable for small laboratories in which the daily number of assays is not high [21].

In the original ^{13}C-UBT, a test meal is administered prior to testing to slow gastric emptying and maximize substrate distribution within the stomach [5]. Test meals containing fat were used to delay gastric emptying. Recently, citric acid, which lowers duodenal pH, resulting in the reduction of antral motility and relaxation of the gastric fundus (gastric accommodation), has been reported to be an optimal test meal [21]. Citric acid test meals have also been shown to enhance the amount of urea hydrolysis compared with traditional nutrient meals without increasing the rate of false-positive results [24–26]. The citric acid test meal also caused a faster increase in breath $^{13}CO_2$, suggesting that breath collection in 10 min after administration may be sufficient for the diagnosis of *H. pylori* infection [25]. Besides citric acid, other test drinks such as orange or apple juice have been used because of a better taste. However, there is controversy over the diagnostic accuracy of UBT using these test drinks [18, 27–29]. Additionally, some studies have reported comparable accuracy without using citric acid (or other test meals) in the UBT [10, 16, 30–35].

9.2.4 Diagnostic Accuracy of the Urea Breath Test in Various Clinical Settings

The diagnostic accuracy of the UBT is excellent according to previous studies. Cochrane review of noninvasive diagnostic tests for *H. pylori* found that UBT yielded high diagnostic accuracy, while serology and stool antigen tests had lower accuracy in detecting *H. pylori* infection [36]. Recent meta-analysis showed that the sensitivity and specificity were 97% (95% confidence interval [CI]: 96–98%) and 96% (95% CI: 95–97%), respectively, with significant heterogeneity between studies [37]. However, it is impossible to determine a universally proposed cut-off level to diagnose *H. pylori* infection because many factors need to be considered, including the test meal, urea dose, type of urea, or the pre-/post-*H. pylori* eradication statuses [12, 15, 16, 18–20, 24, 26, 28, 31, 33, 35, 38–55]. Previously, the cut-off value for the ^{13}C-UBT was proposed to be 5.0‰ based on the normal distribution of excess $\delta^{13}CO_2$ values for *H. pylori*-naïve individuals [56]. Later studies reported that the cut-off value of the ^{13}C-UBT could be lowered from 3.5‰ to 3.0‰ [57], and the cutoff value was even lower at about 2.5‰ with lower doses of urea, and it was generally accepted as the best cutoff point [15, 16, 18, 24, 26, 28, 31, 33, 38–43, 58, 59].

For the purpose of determining successful *H. pylori* eradication, the cutoff values have been proposed between 1.3‰ and 7.4‰ [1, 2]. However, several studies suggested that a ^{13}C-UBT value between 2.0 ‰ and 5.0 ‰ could fall into the "gray zone" [1, 50, 57]. In a Korean retrospective study, the incidence of false-negative results was very low at ^{13}C-UBT values ranging from 2.0‰ to 2.5‰ (1.5%) [55].

With ^{13}C-UBT cutoff values ranging from 2.5‰ to 10.0‰, the incidence of false-positive results ranged from 6.7% to 77.3%. This is common especially among patients with atrophic gastritis, probably attributable to the presence of non-*H. pylori* urease–producing bacteria [55, 60]. In this case, addition of citric acid, which suppresses the delta over baseline value in uninfected individuals and increases the value in subjects with *H. pylori* infection, may improve the accuracy of the UBT.

False-positive results lead to the erroneous conclusion that *H. pylori* eradication has failed [61]. As mentioned above, false-negative UBT results have been reported to occur in ~40% of individuals taking PPIs [18, 59]. This is in part attributable to the effect of PPIs on increasing intra-gastric pH, which could reduce the *H. pylori* bacterial load, at least temporarily. Also, the increased intra-gastric pH after PPI administration can close *H. pylori* urea channel or UreI, thereby reducing the influx of urea into *H. pylori* urease [1]. Besides the PPI's effect on pH, the known direct anti-*H. pylori* antimicrobial effect of the PPI could result in a direct reduction in bacterial load below the critical threshold of urea hydrolysis required for a positive test, although acidification of the stomach did not prevent false-negative UBT results [59]. At least 3-day discontinuation of PPI is necessary before testing for active *H. pylori* infection, and the discontinuation for 14 days is preferred [21].

Another consideration is performing UBT in patients who have undergone partial gastrectomy. After partial gastrectomy, dramatic changes in gastric anatomy and the resultant intragastric environment occurs, with urea passing the stomach faster or slower (due to vagotomy), providing different reactant percentages in the reaction time, or ^{13}C-UBT results may be affected by bile acid reflux. In a hypochlorhydric state of the remnant stomach, intra-gastric overgrowth of non-*H. pylori* urease-producing bacteria (*Streptococcus*, *Staphylococcus*, *Gardnerella*, *Lactococcus*, and *Enterococcus*) can occur [62]. Furthermore, an increase in intra-gastric pH reduces the need for neutralization of hydrochloric acid by bicarbonate and ammonia production by *H. pylori* urease [63–65]. Several studies have reported that ^{13}C-UBT yielded lower diagnostic accuracy when using histology as a reference in the remnant stomach after partial gastrectomy.

The UBT has shown heterogeneous accuracy in the pediatric population, especially in young children, with values of sensitivity and specificity ranging from 75% to 100%, before and after *H. pylori* treatment (using several protocols), despite being a simple and safe noninvasive test in children older than 6 years old [66]. Thus, performance criteria are not yet sufficiently established in children [67].

9.3 Conclusions

The ^{13}C-UBT is an ideal noninvasive *H. pylori* test for those in whom endoscopy is not required. The ^{13}C-UBT offers the combination of simplicity, accuracy, absence of exposure to radioactivity, and reliability, thus the test is widely used in clinical practice, especially to confirm *H. pylori* eradication. However, more studies are necessary to provide standardized protocols and proposed cutoff values of the ^{13}C-UBT in variable clinical conditions.

9.4 Summary

- The urea breath test (UBT) is based on *H. pylori* urease activity, and the ingestion of labeled urea results in the production of labeled carbon dioxide which then can be detected in the breath.
- Urea can be labeled with two different isotopes, ^{14}C (the radioactive isotope) or ^{13}C (the nonradioactive stable isotope). Although ^{14}C-labeled urea is inexpensive, ^{13}C-labeled urea is more stable and nonradioactive, so it is widely used in clinical practice.
- The test protocol is not completely standardized. Factors such as urea dose, whether or not fasting before test, whether a test meal is used, and type of the test meal can affect test accuracy. In general, fasting for >4 h before the test is recommended, and breath sampling is performed 20–30 min after ingestion of the urea

substrate. In general, the use of citric acid test meal is recommended.

- Nevertheless, UBT is a noninvasive, simple, and safe test which provides excellent accuracy both for the diagnosis of *Helicobacter pylori* infection and for the confirmation of successful *H. pylori* eradication.

References

1. Graham DY, Klein PD. Accurate diagnosis of *Helicobacter pylori*. 13C-urea breath test. Gastroenterol Clin North Am. 2000;29(4):885–93. x
2. Gisbert JP, Pajares JM. Review article: 13C-urea breath test in the diagnosis of *Helicobacter pylori* infection – a critical review. Aliment Pharmacol Ther. 2004;20(10):1001–17.
3. Bazzoli F, et al. Urea breath tests for the detection of *Helicobacter pylori* infection. Helicobacter. 1997;2(Suppl 1):S34–7.
4. Thomas JE. 13C urea breath test. Gut. 1998;43(Suppl 3):S7–12.
5. Graham DY, et al. Campylobacter pylori detected noninvasively by the 13C-urea breath test. Lancet. 1987;1(8543):1174–7.
6. Cutler AF, et al. Accuracy of invasive and noninvasive tests to diagnose *Helicobacter pylori* infection. Gastroenterology. 1995;109(1):136–41.
7. Dunn BE. Pathogenic mechanisms of *Helicobacter pylori*. Gastroenterol Clin North Am. 1993;22(1):43–57.
8. Felz MW, Burke GJ, Schuman BM. Breath test diagnosis of *Helicobacter pylori* in peptic ulcer disease: a noninvasive primary care option. J Am Board Fam Pract. 1997;10(6):385–9.
9. Malfertheiner P, et al. Management of *Helicobacter pylori* infection: the Maastricht VI/Florence consensus report. Gut. 2022.
10. Perri F, et al. The influence of isolated doses of drugs, feeding and colonic bacterial ureolysis on urea breath test results. Aliment Pharmacol Ther. 1995;9(6):705–9.
11. Moayyedi P, et al. Do patients need to fast for a 13C-urea breath test? Eur J Gastroenterol Hepatol. 1997;9(3):275–7.
12. Savarino V, et al. Comparison of isotope ratio mass spectrometry and nondispersive isotope-selective infrared spectroscopy for 13C-urea breath test. Am J Gastroenterol. 1999;94(5):1203–8.
13. Rowland M, et al. Carbon 13-labeled urea breath test for the diagnosis of *Helicobacter pylori* infection in children. J Pediatr. 1997;131(6):815–20.
14. Hamlet AK, et al. A simple, rapid, and highly reliable capsule-based 14C urea breath test for diagnosis of *Helicobacter pylori* infection. Scand J Gastroenterol. 1995;30(11):1058–63.
15. Gatta L, et al. A rapid, low-dose, 13C-urea tablet for the detection of *Helicobacter pylori* infection before and after treatment. Aliment Pharmacol Ther. 2003;17(6):793–8.
16. Malaty HM, et al. Twenty-minute fasting version of the US 13C-urea breath test for the diagnosis of *H. pylori* infection. Helicobacter. 1996;1(3):165–7.
17. Bielanski W, Konturek SJ. New approach to 13C-urea breath test: capsule-based modification with low-dose of 13C-urea in the diagnosis of *Helicobacter pylori* infection. J Physiol Pharmacol. 1996;47(3):545–53.
18. Gatta L, et al. Effect of proton pump inhibitors and antacid therapy on 13C urea breath tests and stool test for *Helicobacter pylori* infection. Am J Gastroenterol. 2004;99(5):823–9.
19. Wong WM, et al. Twenty-minute 50 mg 13C-urea breath test without test meal for the diagnosis of *Helicobacter pylori* infection in Chinese. Aliment Pharmacol Ther. 2001;15(9):1499–504.
20. Wong WM, et al. A rapid-release 50-mg tablet-based 13C-urea breath test for the diagnosis of *Helicobacter pylori* infection. Aliment Pharmacol Ther. 2003;17(2):253–7.
21. Savarino V, Vigneri S, Celle G. The 13C urea breath test in the diagnosis of *Helicobacter pylori* infection. Gut. 1999;45(Suppl 1):I18–22.
22. Peura DA, et al. Microdose 14C-urea breath test offers diagnosis of *Helicobacter pylori* in 10 minutes. Am J Gastroenterol. 1996;91(2):233–8.
23. Parente F, Bianchi Porro G. The (13)C-urea breath test for non-invasive diagnosis of *Helicobacter pylori* infection: which procedure and which measuring equipment? Eur J Gastroenterol Hepatol. 2001;13(7):803–6.
24. Klein PD, et al. Noninvasive detection of *Helicobacter pylori* infection in clinical practice: the 13C urea breath test. Am J Gastroenterol. 1996;91(4):690–4.
25. Graham DY, et al. Citric acid as the test meal for the 13C-urea breath test. Am J Gastroenterol. 1999;94(5):1214–7.
26. Dominguez-Munoz JE, et al. A citric acid solution is an optimal test drink in the 13C-urea breath test for the diagnosis of *Helicobacter pylori* infection. Gut. 1997;40(4):459–62.
27. Coelho LG, et al. Application of isotope-selective non-dispersive infrared spectrometry for the evaluation of the 13C-urea breath test: comparison with three concordant methods. Braz J Med Biol Res. 1999;32(12):1493–7.
28. Leodolter A, et al. Citric acid or orange juice for the 13C-urea breath test: the impact of pH and gastric emptying. Aliment Pharmacol Ther. 1999;13(8):1057–62.
29. Labenz J, et al. Validity of a novel biopsy urease test (HUT) and a simplified 13C-urea breath test for diagnosis of *Helicobacter pylori* infection and estimation of the severity of gastritis. Digestion. 1996;57(6):391–7.

30. Chen TS, et al. Simplified 13C-urea breath test with a new infrared spectrometer for diagnosis of *Helicobacter pylori* infection. J Gastroenterol Hepatol. 2003;18(11):1237–43.
31. Gisbert JP, et al. 13C-urea breath test for the diagnosis of *Helicobacter pylori* infection before treatment: is citric acid necessary? Dig Liver Dis. 2000;32(1):20–4.
32. Oksanen A, et al. Accurate detection of *Helicobacter pylori* infection with a simplified 13C urea breath test. Scand J Clin Lab Invest. 1997;57(8):689–94.
33. Miwa H, et al. Usefulness of the [13C]-urea breath test for detection of *Helicobacter pylori* infection in fasting patients. J Gastroenterol Hepatol. 1998;13(10):1039–43.
34. Casellas F, et al. The impact of delaying gastric emptying by either meal substrate or drug on the [13C]-urea breath test. Am J Gastroenterol. 1999;94(2):369–73.
35. Wong WM, et al. (13)C-urea breath test without a test meal is highly accurate for the detection of *Helicobacter pylori* infection in Chinese. Aliment Pharmacol Ther. 2000;14(10):1353–8.
36. Best LM, et al. Non-invasive diagnostic tests for *Helicobacter pylori* infection. Cochrane Database Syst Rev. 2018;3(3):Cd012080.
37. Abd Rahim MA, et al. (13)C-Urea breath test accuracy for *Helicobacter pylori* infection in the Asian population: a meta-analysis. Ann Glob Health. 2019;85(1).
38. Braden B, et al. More convenient 13C-urea breath test modifications still meet the criteria for valid diagnosis of *Helicobacter pylori* infection. Z Gastroenterol. 1994;32(4):198–202.
39. Gisbert JP, et al. Validation of the 13c-urea breath test for the initial diagnosis of *Helicobacter pylori* infection and to confirm eradication after treatment. Rev Esp Enferm Dig. 2003;95(2):121–6. 115–20
40. Hamlet A, et al. A novel tablet-based 13C urea breath test for *Helicobacter pylori* with enhanced performance during acid suppression therapy. Scand J Gastroenterol. 1999;34(4):367–74.
41. Kato M, et al. Ten minute 13C-urea breath test for the diagnosis of *Helicobacter pylori* infection. J Gastroenterol. 1998;33(Suppl 10):40–3.
42. Leodolter A, et al. Validity of a modified 13C-urea breath test for pre- and posttreatment diagnosis of *Helicobacter pylori* infection in the routine clinical setting. Am J Gastroenterol. 1999;94(8):2100–4.
43. Liao CC, et al. The 13C-urea breath test to detect *Helicobacter pylori* infection: a validated simple methodology with 50 mg 13C-urea. Aliment Pharmacol Ther. 2002;16(4):787–92.
44. Ng FH, et al. [13C]-urea breath test without prior fasting and without test meal is accurate for the detection of *Helicobacter pylori* infection in Chinese. J Gastroenterol Hepatol. 2002;17(8):834–8.
45. Savarino V, et al. Isotope ratio mass spectrometry (IRMS) versus laser-assisted ratio analyzer (LARA): a comparative study using two doses of. Dig Dis Sci. 2000;45(11):2168–74.
46. Sheu BS, et al. Lower-dose (13)C-urea breath test to detect *Helicobacter pylori* infection-comparison between infrared spectrometer and mass spectrometry analysis. Aliment Pharmacol Ther. 2000;14(10):1359–63.
47. Sheu BS, et al. Selection of lower cutoff point of [13C]urea breath test is helpful to monitor *H. pylori* eradication after proton pump inhibitor-based triple therapy. Dig Dis Sci. 2000;45(7):1330–6.
48. Gomollón F, et al. Breath test is very reliable for diagnosis of *Helicobacter pylori* infection in real clinical practice. Dig Liver Dis. 2003;35(9):612–8.
49. Urita Y, et al. Breath sample collection through the nostril reduces false-positive results of 13C-urea breath test for the diagnosis of *Helicobacter pylori* infection. Dig Liver Dis. 2004;36(10):661–5.
50. Calvet X, et al. Accuracy of diagnostic tests for *Helicobacter pylori*: a reappraisal. Clin Infect Dis. 2009;48(10):1385–91.
51. Peng NJ, et al. Comparison of noninvasive diagnostic tests for *Helicobacter pylori* infection. Med Princ Pract. 2009;18(1):57–61.
52. Peng NJ, et al. Capsule 13C-urea breath test for the diagnosis of *Helicobacter pylori* infection. World J Gastroenterol. 2005;11(9):1361–4.
53. Wardi J, et al. A rapid continuous-real-time 13C-urea breath test for the detection of *Helicobacter pylori* in patients after partial gastrectomy. J Clin Gastroenterol. 2012;46(4):293–6.
54. Li ZX, et al. Cut-off optimization for (13)C-urea breath test in a community-based trial by mathematic, histology and serology approach. Sci Rep. 2017;7(1):2072.
55. Kwon YH, et al. The diagnostic validity of citric acid-free, high dose (13)C-urea breath test after *Helicobacter pylori* eradication in Korea. Helicobacter. 2015;20(3):159–68.
56. Logan RP. Urea breath tests in the management of *Helicobacter pylori* infection. Gut. 1998;43(Suppl 1):S47–50.
57. Mion F, et al. 13C-urea breath test for *Helicobacter pylori*: cut-off point determination by cluster analysis. Clin Sci (Lond). 1997;93(1):3–6.
58. Kopacova M, et al. Comparison of different protocols for 13C-urea breath test for the diagnosis of *Helicobacter pylori* infection in healthy volunteers. Scand J Clin Lab Invest. 2005;65(6):491–8.
59. Graham DY, et al. Studies regarding the mechanism of false negative urea breath tests with proton pump inhibitors. Am J Gastroenterol. 2003;98(5):1005–9.
60. Graham DY, et al. Simplified 13C-urea breath test for detection of *Helicobacter pylori* infection. Am J Gastroenterol. 2001;96(6):1741–5.
61. Graham DY. *Helicobacter pylori* update: gastric cancer, reliable therapy, and possible benefits. Gastroenterology. 2015;148(4):719–31.e3.
62. Brandi G, et al. Urease-positive bacteria other than *Helicobacter pylori* in human gastric juice and mucosa. Am J Gastroenterol. 2006;101(8):1756–61.
63. Bell NJ, Hunt RH. Progress with proton pump inhibition. Yale J Biol Med. 1992;65(6):649–57.

64. Gurbuz AK, et al. Is the remarkable contradiction between histology and 14C urea breath test in the detection of *Helicobacter pylori* due to false-negative histology or false-positive 14C urea breath test? J Int Med Res. 2005;33(6):632–40.
65. Osaki T, et al. Urease-positive bacteria in the stomach induce a false-positive reaction in a urea breath test for diagnosis of *Helicobacter pylori* infection. J Med Microbiol. 2008;57(Pt 7):814–9.
66. Guarner J, et al. *Helicobacter pylori* diagnostic tests in children: review of the literature from 1999 to 2009. Eur J Pediatr. 2010;169(1):15–25.
67. Machado RS, Patricio FR, Kawakami E. 13C-urea breath test to diagnose *Helicobacter pylori* infection in children aged up to 6 years. Helicobacter. 2004;9(1):39–45.

10 Other Diagnostic Methods

Jung-Hwan Oh

Highlights

What Is Known
• *Helicobacter pylori* (*H. pylori*) can be diagnosed by invasive biopsy or noninvasive methods such as urea breath testing, serology, and stool antigen testing (SAT). • SATs, particularly those using monoclonal antibodies, have high sensitivity and specificity for both initial diagnosis and confirmation after eradication therapy. • Noninvasive tests are advantageous because they assess the whole stomach and do not require an overnight fast. • Although saliva-based PCR methods for *H. pylori* detection have been studied, they are less reliable and not widely used in clinical practice.
Future Research Questions
• Further studies are warranted to develop new noninvasive methods and more accurate tests.

10.1 Introduction

The initial test for detecting *Helicobacter pylori* is an invasive biopsy obtained during upper gastrointestinal endoscopy. However, a diagnosis of *H. pylori* can be achieved through noninvasive methods such as urea breath testing (UBT), serology, and stool antigen test (SAT). These noninvasive tests prove valuable in situations where upper gastrointestinal endoscopy is unavailable or for patients who do not necessitate endoscopic evaluation. While UBT and serologic tests have already been discussed, the focus here will be on SAT.

J.-H. Oh (✉)
Department of Internal Medicine, Eunpyeong St. Mary's Hospital, The Catholic University of Korea, Seoul, Korea
e-mail: ojh@catholic.ac.kr

10.2 Main Body

10.2.1 Stool Antigen Test

There are two types of antibody tests: polyclonal and monoclonal. In a study comparing the accuracy of monoclonal and multiclonal SATs, the results of 13 studies indicated a higher integrated sensitivity for monoclonal techniques (0.95 vs. 0.83) [1]. Another study demonstrated that the sensitivity and specificity of the monoclonal SAT were 97% and 94%, respectively, suggesting it as an excellent alternative among non-invasive tests [2]. Therefore, monoclonal antibody tests are recognized for their superior sensitivity, specificity, and overall accuracy compared to polyclonal antibody tests. Due to low sensitivity, SATs employing polyclonal antibody tests are not commonly utilized [3].

SAT can serve both for confirming the initial diagnosis of *H. pylori* and assessing the success of eradication therapy [4]. In a meta-analysis comparing 22 studies (including approximately

B.-W. Kim (ed.), *Helicobacter Infection in Clinical Practice*,
https://doi.org/10.1007/978-981-97-8397-7_10

2500 individuals) focusing on the monoclonal SAT for the initial diagnosis before treatment of *H. pylori*, the sensitivity and specificity were found to be 0.94 (95% CI 0.93–0.95) and 0.97 (0.96–0.98), respectively [1]. Additionally, a Japanese study investigating the utility of monoclonal antibody-based SAT for confirming eradication therapy reported a concordance rate of more than 90% when compared to UBT. This suggests that SAT, even when employing monoclonal antibodies, can be effectively utilized for confirming eradication therapy, comparable to the reliability of UBT [5].

Twelve studies, involving 957 patients, assessed the effectiveness of monoclonal SATs in determining treatment outcomes. The combined sensitivity and specificity were determined to be 0.93 (0.89–0.96) and 0.96 (0.94–0.97), respectively [1]. PCR-based tests play a valuable role in clinical practice by not only providing diagnostic information but also insights into antibiotic resistance [6]. Innovative approaches, such as the combination of easyMAG(®) (bioMérieux) with Amplidiag® H pylori + ClariR (Mobidiag), are showing promise [7]. It is advisable to wait approximately 3 months after completing treatment to confirm eradication [8].

Confirmation of eradication in all patients treated for *H. pylori* is recommended through noninvasive tests (UBT, SAT) or invasive test like endoscopy-based test. Noninvasive test offers the advantage of evaluating the entire stomach contents, while invasive test using a biopsy may assess only a portion of the stomach, with results possibly varying based on the endoscopist. Additionally, SAT has an advantage over UBT in that it does not require fasting like the UBT. Therefore, noninvasive tests such as UBT and SAT are considered optimal for detecting active *H. pylori* infection, although recent intake of medications (such as proton-pump inhibitors, potassium competitive acid blockers, bismuth-containing compounds, or antibiotics) may lead to false-negative results.

H. pylori is known to spread through the fecal-oral route, prompting researchers to explore diagnostic methods in both saliva as well as stool. Studies have investigated a PCR-based approach for detecting *H. pylori* DNA in saliva samples. In a Thai study, *H. pylori* was identified in both saliva and fecal samples in 64% of healthy adults [9]. However, variations in *H. pylori* DNA detection rates between saliva and stool have been observed in other studies [10]. The diagnosis of *H. pylori* infection through saliva is considered less reliable, as indicated by a study reporting significant differences [11]. While saliva-based diagnosis offers convenience in clinical practice, its limited reliability has hindered universal adoption.

10.3 Conclusion

Various methods for detecting *H. pylori* are described in this chapter. The optimal method can be chosen based on the physician's preference and the situation.

References

1. Gisbert JP, de la Morena F, Abraira V. Accuracy of monoclonal stool antigen test for the diagnosis of H. pylori infection: a systematic review and meta-analysis. Am J Gastroenterol. 2006;101:1921–30.
2. Dore MP, Negrini R, Tadeu V, et al. Novel monoclonal antibody-based Helicobacter pylori stool antigen test. Helicobacter. 2004;9:228–32.
3. Korkmaz H, Findik D, Ugurluoglu C, Terzi Y. Reliability of stool antigen tests: investigation of the diagnostic value of a new immunochromatographic Helicobacter pylori approach in dyspeptic patients. Asian Pac J Cancer Prev. 2015;16:657–60.
4. Chey WD, Wong BC. American College of Gastroenterology guideline on the management of Helicobacter pylori infection. Am J Gastroenterol. 2007;102:1808–25.
5. Shimoyama T, Kobayashi I, Kato C, Kodama M, Fukuda Y. Comparison of monoclonal antibody-based stool antigen tests to determine the results of Helicobacter pylori eradication therapy. Scand J Gastroenterol. 2010;45:1431–4.
6. Patel SK, Pratap CB, Jain AK, Gulati AK, Nath G. Diagnosis of Helicobacter pylori: what should be the gold standard? World J Gastroenterol. 2014;20:12847–59.
7. Godbole G, Mégraud F, Bessède E. Review: Diagnosis of Helicobacter pylori infection. Helicobacter. 2020;25(Suppl 1):e12735.
8. Koletzko S, Konstantopoulos N, Bosman D, et al. Evaluation of a novel monoclonal enzyme immuno-

assay for detection of Helicobacter pylori antigen in stool from children. Gut. 2003;52:804–6.
9. Wongphutorn P, Chomvarin C, Sripa B, Namwat W, Faksri K. Detection and genotyping of Helicobacter pylori in saliva versus stool samples from asymptomatic individuals in Northeastern Thailand reveals intra-host tissue-specific H. pylori subtypes. BMC Microbiol. 2018;18:10.
10. Kabir S. Detection of Helicobacter pylori DNA in feces and saliva by polymerase chain reaction: a review. Helicobacter. 2004;9:115–23.
11. Berloco P, Cavallini A, Di Leo A, Russo F. Saliva samples not a reliable tool for diagnosis of Helicobacter pylori infection. Eur J Clin Microbiol Infect Dis. 2001;20:68–9.

Culture-Based Antibiotic Resistance Tests

11

Ji Yong Ahn

Highlights

What Is Known
• Antibiotic susceptibility test before *H. pylori* eradication can increase success rate of treatment and, ultimately, can decrease the antibiotic-resistant of *H. pylori*.
Future Research Questions
• To use more tailored therapy for *H. pylori* eradication, we should try to find more effective and convenient methods of antibiotic susceptibility test about *H. pylori*.

11.1 Introduction

The eradication rate of standard triple therapy, comprising clarithromycin, amoxicillin, and proton-pump inhibitor, widely used for *H. pylori* eradication, has been declining. Therefore, the eradication rate is reported to be approximately 70–80% and below 70% in 7-day standard triple therapy [1, 2]. This reduction is attributed to increased antibiotic resistance in *H. pylori*, especially in clarithromycin which is most effective and commonly utilized as a primary eradication regimen [3, 4]. Besides clarithromycin, other antibiotics utilized for *H. pylori* eradication such as metronidazole, amoxicillin, tetracycline, and quinolone also have shown rising resistance [5]. Moreover, multidrug resistance involving two or more antibiotics is a contemporary challenge, further impacting treatment success rate [5].

To overcome the antibiotic resistance, numerous trials explore strategies like adding bismuth, using multiple antibiotics, extending treatment duration, and increasing antibiotic dosages [6, 7]. While the optimal approach would involve determining antibiotics susceptibility before treatment, the standard method, agar dilution test, requires isolating viable *H. pylori*, demanding specific microbiology skills in the clinical laboratory to grow this fastidious microaerophilic bacterium over 1–2 weeks. However, accumulating knowledge in some centers and the development of polymerase chain reaction kits or sequencing methods, detecting 23S ribosomal RNA point mutations related to clarithromycin resistance, have revived interest in antibiotic susceptibility test (AST). This approach aims to identify effective antibiotics, avoid ineffective treatments, minimize antibiotic resistance, and tailor treatment plans for individuals. This chapter outlines the antibiotic resistance test method for *H. pylori* and treatment outcomes based on the AST.

J. Y. Ahn (✉)
Department of Gastroenterology, Asan Medical Center, University of Ulsan College of Medicine, Seoul, Korea

B.-W. Kim (ed.), *Helicobacter Infection in Clinical Practice*,
https://doi.org/10.1007/978-981-97-8397-7_11

11.2 Main Body

11.2.1 The Standard Antimicrobial Susceptibility Test of *H. pylori*

The Clinical and Laboratory Standards Institute (CLSI) recommends the agar dilution method as standard susceptibility testing of *H. pylori*, and set the minimal inhibitory concentration (MIC) ≥1.0 mg/L as clarithromycin resistance breakpoint [8]. Until now, the clinical MIC breakpoints for other anti-*H. pylori* antibiotics including amoxicillin, metronidazole, levofloxacin, tetracycline, and rifabutin have not yet been established, so some institute made MIC breakpoints for these antibiotics by their own data [9].

Agar dilution susceptibility testing has been considered as the "gold standard" to determine the MICs of antimicrobial agents and also as the reference method to compare other AST methods [10]. This method involves the incorporation of different concentrations of antimicrobials into blood agar plates [8], therefore the need to prepare many freshly made blood agar plates for agar dilution susceptibility testing is both labor-intensive and time-consuming work.

Usual methods are as follows: two mucosal biopsy specimens are obtained from gastric antrum and corpus greater curvature, using standard-sized biopsy forceps during endoscopy. Samples are transported to the laboratory and maintained at room temperature, without transport media, until processing. Specimens are trimmed with a razor and streaked onto Brucella agar supplemented with 7% defibrinated sheep blood containing vancomycin (10 μg/mL), trimethoprim (5 μg/mL), amphotericin B (5 μg/mL), and polymyxin B (2.5 IU). The plates are incubated at 37 °C, under microaerophilic conditions for 5–7 days. After the initial isolation, colonies from each biopsy specimen are subcultured and identified as *H. pylori* based on their morphology and positive reactions with urease, oxidase, and catalase. All stock cultures are stored in Brucella broth supplemented with 15% glycerol and 10% fetal bovine serum at −80 °C. These preparations are thawed and subcultured for subsequent experiments. The MIC of clarithromycin against *H. pylori* strain is determined by using the serial twofold agar dilution method. The MIC value is defined at a minimum dilution concentration of the antimicrobial agent that does not produce bacterial colonies, and the resistant breakpoint of MIC is set at >0.5 μg/mL for clarithromycin [9, 11].

The detailed methods of antibiotic resistance test are described as follows.

11.2.1.1 *H. pylori* Culture

1. Test medium: 7% Brucella blood agar plate [7% sheep blood, Vancomycin (10 μg/mL), Trimethoprim (5 μg/mL), Amphotericin B (5 μg/mL), Polymyxin B (2.5 IU)]
2. Spread the gastric tissue stored in a deep freezer at −80 °C on a culture test medium using a sterile loop.
3. The culture medium inoculated with *H. pylori* strain is placed in a humidified CO_2 incubator and incubated at 37 °C and 10% CO_2 for 4–7 days.
4. If a colony of *H. pylori* is identified, separate the single colony and spread it in a new medium and incubate it for 3–4 days (Fig. 11.1).
5. The cultured *H. pylori* is collected in a sterile loop and spread in a new medium and incubated for 3–4 days.
6. Collect strains and place them in a 15% glycerol/tryptic soy broth (TSB) medium and store them in a deep freezer at −70 °C.

11.2.1.2 Biological Activity Test

1. Check the *H. pylori* based on the urease test, oxidase test, catalase test, and UreC PCR methods.
2. Urease test: To check the urease activity of *H. pylori*, put a part of the strain into the urea broth to check the color change.
3. Catalase test: Add strains to 3% H_2O_2 to check for bubbles.
4. Oxidase test: Absorb strains into an oxidase test disk containing an oxidase agent to check for color changes.

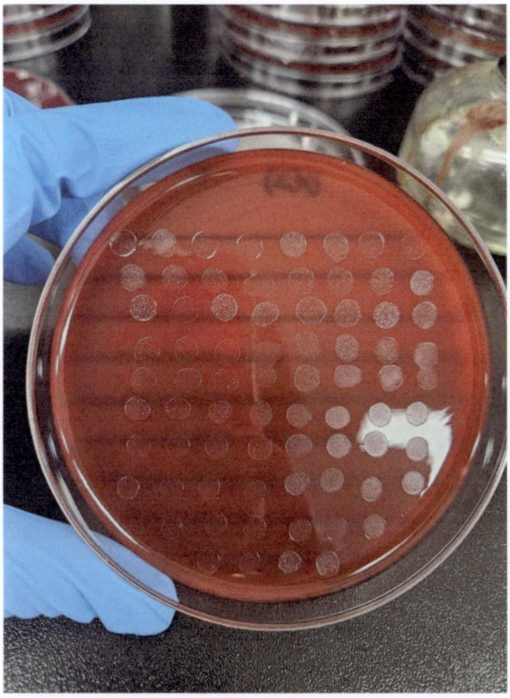

Fig. 11.1 Colony of *Helicobacter pylori*

5. UreC PCR: To check the UreC gene of *H. pylori*, amplify it with PCR and check the size of the PCR product by electrophoresis.

11.2.1.3 Antibiotic Susceptibility Test

1. The agar dilution method recommended by the Clinical and Laboratory Standards Institute (CLSI) is used to test the antibiotic sensitivity of *H. pylori* (Fig. 11.2).
2. Test medium: Mueller Hinton agar medium with 5% sheep blood and diluted antibiotics.
3. Dilution of antibiotics: Prepare the antibiotics (amoxicillin, clarithromycin, metronidazole, tetracycline, ciprofloxacin, and levofloxacin) for testing by diluting them twice in a row with sterilized distilled water.
4. Dilution concentration of antibiotics: Dilution concentration is set from 0.008 to 256 μg/mL, and the meaning of the dilution concentration varies by antibiotic.
5. The *H. pylori* strain cultured for 48 h in a *H. pylori* culture medium is mixed with PBS, adjusted to McFarland=2 using a densitometer, and inoculated 5 μL into the antibiotic susceptibility test medium prepared for each antibiotic concentration.
6. Confirmation of bacterial colony formation: After incubating the test medium inoculated with *H. pylori* strain at 37 °C and 10% CO_2 for 72 h, observe whether the bacterial colony is formed.

11.2.1.4 Criteria and Evaluation Methods

1. Check the antibiotic dilution concentration (MIC) at which *H. pylori* bacterial colonies.
2. Evaluate the adequacy of antibiotic susceptibility tests in the MIC range of standard strain ATCC 43504.

11.2.2 Clinical Outcomes of Tailored Therapy

Guidelines recommend conducting AST at least before the third line of *H. pylori* eradication. If endoscopic examination is available, considering culture-based treatment is recommended even in the first and second therapy [12–15]. A recent systematic review of comparative studies demonstrated that *H. pylori* infection was cured in 89.9% of 2052 patients treated with antibiotic-tailored therapies and in 77.6% of 2516 patients receiving empiric therapy ($P < 0.001$) [16]. In another study based on AST, the eradication rates was 94.7% in the antimicrobial susceptibility-guided group and 71.9% in the clarithromycin-based triple therapy group after the initial treatment ($P = 0.002$) in intension-to-treat analysis, while the per-protocol eradication rates were 96.4% in the antimicrobial susceptibility-guided group and 73.2% in the clarithromycin-based triple therapy group ($P = 0.001$) [17]. Another study comparing the results of standard triple therapy and sequential therapy based on AST showed that eradication rates in the antimicrobial susceptibility-guided group and empirical therapy groups were 85.4% and 58.4% ($P < 0.01$). In standard triple therapy, eradication rates of the antimicrobial susceptibility-guided group and

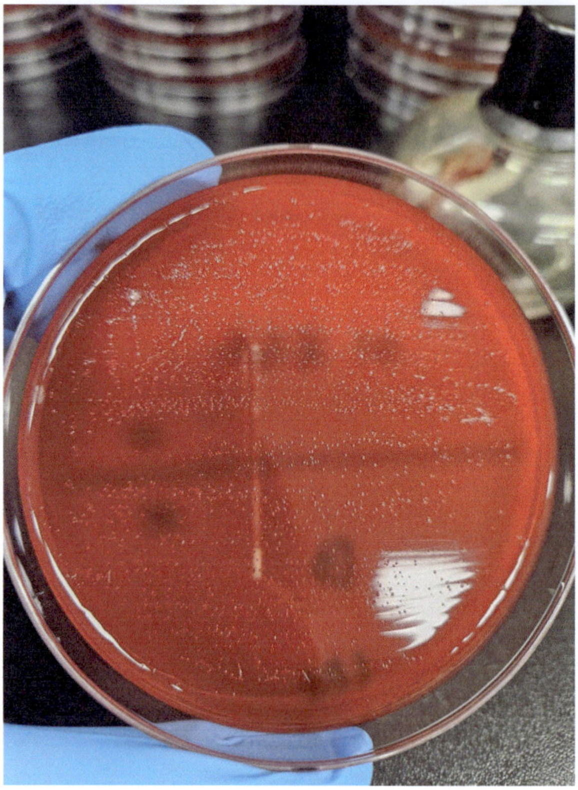

Fig. 11.2 Antibiotic susceptibility test by the Agar dilution method

empirical therapy groups were 85.1% and 56.6% ($P < 0.01$). In sequential therapy, eradication rates of the antimicrobial susceptibility-guided group and empirical therapy groups were 86.2% and 65.6% ($P = 0.06$). According to AST, standard triple therapy had an eradication rate of 84.6% with strains susceptible to clarithromycin and amoxicillin and 11.1% with strains resistant to both. Sequential therapy had an eradication rate of 89.5% with strains susceptible to clarithromycin, amoxicillin, and metronidazole, whereas it was 0% with strains resistant to clarithromycin and metronidazole [18].

11.3 Conclusion

Even though the culture of *H. pylori* and performing AST before eradication is not easy in clinical practice, the consideration of AST before treatment or following treatment failure is crucial. This practice can play a pivotal role in optimizing patient care and mitigating the spread of antibiotic-resistant strains of *H. pylori.*

References

1. De Francesco V, et al. Worldwide *H. pylori* antibiotic resistance: a systematic review. J Gastrointestin Liver Dis. 2010;19(4):409–14.
2. Choi YJ, et al. Triple therapy-based on tegoprazan, a new potassium-competitive acid blocker, for first-line treatment of *Helicobacter pylori* infection: a randomized, double-blind, phase III, clinical trial. Gut Liver. 2022;16(4):535–46.
3. Lee JY, et al. Primary and secondary antibiotic resistance of *Helicobacter pylori* in Korea from 2003 to 2018. Helicobacter. 2019;24(6):e12660.
4. Lee JH, et al. Nationwide antibiotic resistance mapping of *Helicobacter pylori* in Korea: a prospective multicenter study. Helicobacter. 2019;24(4):e12592.
5. Lee JW, et al. Prevalence and trends of multiple antimicrobial resistance of *Helicobacter pylori* in one

tertiary hospital for 20 years in Korea. Helicobacter. 2023;28(1):e12939.
6. Shin DW, et al. The benefit of the bismuth add-on to the 2-week clarithromycin-based triple regimen for *Helicobacter pylori* eradication: a propensity score-matched retrospective study. Gut Pathog. 2023;15(1):13.
7. Leow AH, et al. Optimizing first line 7-day standard triple therapy for *Helicobacter pylori* eradication: prolonging treatment or adding bismuth: which is better? J Dig Dis. 2018;19(11):674–7.
8. Jorgensen JH, Hindler JF. New consensus guidelines from the Clinical and Laboratory Standards Institute for antimicrobial susceptibility testing of infrequently isolated or fastidious bacteria. Clin Infect Dis. 2007;44(2):280–6.
9. Kim JM, et al. Comparison of primary and secondary antimicrobial minimum inhibitory concentrations for *Helicobacter pylori* isolated from Korean patients. Int J Antimicrob Agents. 2006;28(1):6–13.
10. Wiegand I, Hilpert K, Hancock RE. Agar and broth dilution methods to determine the minimal inhibitory concentration (MIC) of antimicrobial substances. Nat Protoc. 2008;3(2):163–75.
11. EUCAST clinical breakpoint tables v. 10.0, valid from 2020-01-01. *Helicobacter pylori*. http://www.eucast.org/clinical_breakpoints. Accessed 17 May 2020.
12. Malfertheiner P, et al. Management of *Helicobacter pylori* infection–the Maastricht IV/ Florence Consensus Report. Gut. 2012;61(5):646–64.
13. Malfertheiner P, et al. Management of *Helicobacter pylori* infection-the Maastricht V/Florence Consensus Report. Gut. 2017;66(1):6–30.
14. Kato M, et al. Guidelines for the management of *Helicobacter pylori* infection in Japan: 2016 Revised Edition. Helicobacter. 2019;24(4):e12597.
15. El-Serag HB, et al. Houston Consensus conference on Testing for *Helicobacter pylori* infection in the United States. Clin Gastroenterol Hepatol. 2018;16(7):992–1002.e6.
16. Francesco V, et al. Culture-based antibiotic susceptibility testing for *Helicobacter pylori* infection: a systematic review. Ann Gastroenterol. 2022;35(2):127–34.
17. Park CS, et al. Pretreatment antimicrobial susceptibility-guided vs. clarithromycin-based triple therapy for *Helicobacter pylori* eradication in a region with high rates of multiple drug resistance. Am J Gastroenterol. 2014;109(10):1595–602.
18. Kang S, et al. Role of antimicrobial susceptibility testing before first-line treatment containing clarithromycin for *Helicobacter pylori* eradication in the clinical setting. Antibiotics (Basel). 2021;10(2)

PCR-Based Antibiotic Resistance Tests

12

Jae Myung Park

Highlights

What Is Known
• The eradication of *Helicobacter pylori* is possible only through the use of a restricted range of antibiotics. • The confirmation of antibiotic resistance is of utmost importance in ensuring the efficacy of *Helicobacter pylori* eradication treatment. • *Helicobacter pylori* has been detected in a biopsy of the stomach, cultivated, and subsequently its resistance to antibiotics was verified through an antibiotic susceptibility test.
Future Research Questions
• In recent times, various molecular biology techniques have been developed to validate the antibiotic resistance of *Helicobacter pylori*. • This chapter will comprehensively explore the cutting-edge information pertaining to this matter.

12.1 Introduction

In the eradication of *Helicobacter pylori* (*H. pylori*), it is crucial to assess antibiotic resistance prior to treatment. Ultimately, the most effective strategy entails devising a customized regimen that specifically targets *H. pylori* with antibiotics that it is sensitive to. This customized treatment can effectively eliminate the organism without engendering undesirable adverse events. The primary modalities for identifying antibiotic resistance are culture-based techniques and genotypic analysis. Both methods necessitate the procurement of gastric mucosal samples through upper gastrointestinal endoscopy.

12.2 Main Body

The empirical approach to eradicating *H. pylori* does not take into account the resistance to antibiotics, resulting in a decline in the success rate of eradication. Notably, among the various antimicrobial resistances, clarithromycin poses a significant concern in the eradication of *H. pylori* due to its essential role in the most frequently employed treatment regimens. In the presence of clarithromycin resistance, the overall eradication rate is approximately 60% when using first-line empirical regimens [1]. Therefore, it is not advisable to utilize clarithromycin as a first-line treatment in regions where there is a high prevalence of antibiotic resistance [2, 3].

The culture-based test is a phenotypic testing method that determines antibiotic susceptibility. Although it provides high specificity, it necessitates specialized facilities, advanced techniques, and labor-intensive efforts. In addition, the success rate is not considerably high, and the process consumes a significant amount of time,

J. M. Park (✉)
Department of Internal Medicine, Seoul St. Mary's Hospital, The Catholic University of Korea, Seoul, Korea

B.-W. Kim (ed.), *Helicobacter Infection in Clinical Practice*,
https://doi.org/10.1007/978-981-97-8397-7_12

which is necessary to obtain information on antibiotic resistance. Consequently, the extensive implementation in clinical practice poses challenges [4].

The primary approach for diagnosing clarithromycin resistance was through culture-based methods, which involved measuring the minimal inhibitory concentration (MIC). This concentration refers to the lowest amount of antibiotics that can effectively hinder the proliferation of microorganisms. The reported MIC for *H. pylori* antibiotics shows significant variation due to differences in strains across regions, diverse culture conditions, low sensitivity, and inconsistent test methods employed by different entities.

By utilizing polymerase chain reaction (PCR)-based techniques, genotypic analysis can effectively pinpoint specific regions of the *H. pylori* genome DNA. PCR, including real-time PCR, restriction fragment length polymorphism PCR, and nested PCR, is a cost-effective, precise, and fast technology for detecting *H. pylori* infection, even in cases of low bacterial load or simultaneous use of proton-pump inhibitors (PPIs). A variety of genes, including 16S rRNA, 23S rRNA, *cagA*, *flaA*, *vacA*, and *ure* genes, have been considered in the process of designing primers.

The PCR technique is also capable of identifying point mutations that are associated with antibiotic resistance. It is worth noting that there is a strong correlation between genetic resistance to clarithromycin in *H. pylori* and the resistance results obtained from culture tests. As a result, genetic mutation testing can be seen as a feasible alternative to *H. pylori* culture and antibiotic susceptibility test, delivering prompt, convenient, and accurate resistance results [5].

Clinical studies demonstrate that individualized treatment, relying on the identification of the clarithromycin resistance gene, yields a superior eradication rate in contrast to the use of empirical treatment [6–8]. The PCR-based method has become a standard practice in identifying various infectious agents, including viruses and bacteria. When diagnosing *H. pylori* infection, the PCR-based method exhibits relatively high sensitivity (82~90%) and specificity (95% or greater) [9–11]. Furthermore, it provides the advantage of concurrently assessing resistance to antibiotics such as clarithromycin, metronidazole, and quinolone, which is restricted in culture tests.

Nevertheless, there are certain limitations, such as the occurrence of false positives attributable to the presence of nonviable *H. pylori* genetic material and false negatives arising from sampling errors, the expensive nature of the equipment, and the necessity for additional facilities. With the aim of minimizing the occurrence of false positives, multiplex PCR has been implemented, employing two primers of disparate lengths connected by a polydeoxyinosine linker [12]. According to a study, this new PCR test showcases a sensitivity of 87.5% and a specificity of 91.3%, accompanied by positive and negative predictive rates of 84.0% and 93.3%, and an accuracy of 90.0% [8].

The main cause of clarithromycin resistance in *H. pylori* is attributed to point mutations occurring at position 2142 or 2143 of the 23S rRNA gene. The primary point mutation takes place when adenine is substituted with guanine at positions 2142 and 2143 (A2142G and A2143G, respectively; Fig. 12.1). Infrequent occurrences consist of adenine substitution with cytosine at position 2142 (A2142C), along with A2115G, G2141A, A2142T, or T2182C substitutions [13, 14]. The regional distribution of these mutations varies.

In the United States, the A2142G and A2143G mutations account for 48–53% and 39–45% of cases, respectively. In Europe, the A2142G and A2143G mutations are reported to occur in 23–33% and 44–67% of cases, respectively [13, 15, 16]. However, in Eastern countries, the A2143G mutation constituted over 90% of cases [6, 17–19]. It is noteworthy that the A2143G mutation is predominantly associated with clarithromycin resistance cases [20–22].

Additional genotypic analyses comprise real-time PCR, nested PCR, and multiplex PCR. Real-time PCR, which stands out due to its quantitative, rapid, and sensitive attributes, proves to be highly effective in the testing of *H. pylori* [23, 24]. Typically, nested PCR shows more sensitivity than conventional PCR [25–28]. Multiplex PCR, on the other hand, is specifi-

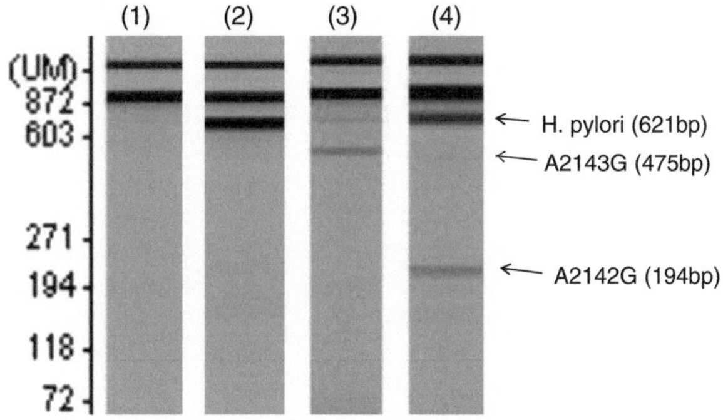

Fig. 12.1 A demonstration of *Helicobacter pylori* detection and evaluation of clarithromycin antibiotic resistance. (1) *Helicobacter* negative, (2) *Helicobacter* positive with no clarithromycin resistance, (3) *Helicobacter* is positive and is accompanied by the clarithromycin-related genetic code (A2143G), (4) *Helicobacter* is positive and is accompanied by the clarithromycin-related genetic code (A2142G)

cally designed to identify multiple target sequences. In the case of *H. pylori*, this method facilitates the concurrent identification of *vacA* alleles, *cagA* gene, and antibiotic resistance in a single reaction [29, 30].

12.3 Conclusion

In summary, the emergence of clarithromycin resistance plays a significant role in the decline of *H. pylori* eradication rates. Genetic tests present a method to streamline the process, thereby reducing both the time and costs linked to multiple tests. While uncertainties may arise regarding the comprehensive identification of clarithromycin resistance through this system and the selection of the most efficacious drug, continuous advancements in genetic testing hold promise for enhancing eradication regimens in the future.

References

1. Zou Y, et al. The effect of antibiotic resistance on *Helicobacter pylori* eradication efficacy: a systematic review and meta-analysis. Helicobacter. 2020;25(4):e12714.
2. Graham DY, Lee YC, Wu MS. Rational *Helicobacter pylori* therapy: evidence-based medicine rather than medicine-based evidence. Clin Gastroenterol Hepatol. 2014;12(2):177–86.e3. Discussion e12-3
3. Graham DY, Liou JM. Primer for development of guidelines for *Helicobacter pylori* therapy using antimicrobial stewardship. Clin Gastroenterol Hepatol. 2022;20(5):973–983.e1.
4. Cammarota G, et al. Culture-guided treatment approach for *Helicobacter pylori* infection: review of the literature. World J Gastroenterol. 2014;20(18):5205–11.
5. Wang YH, et al. A systematic review and meta-analysis of genotypic methods for detecting antibiotic resistance in *Helicobacter pylori*. Helicobacter. 2018;23(2):e12467.
6. Kim SY, et al. Types of 23S ribosomal RNA point mutations and therapeutic outcomes for *Helicobacter pylori*. Gut Liver. 2021;15(4):528–36.
7. Kwon YH, et al. Efficacy of tailored therapy for *Helicobacter pylori* eradication based on clarithromycin resistance and survey of previous antibiotic exposure: a single-center prospective pilot study. Helicobacter. 2019;24(4):e12585.
8. Lee HJ, et al. Eradication of *Helicobacter pylori* according to 23S ribosomal RNA point mutations associated with clarithromycin resistance. J Infect Dis. 2013;208(7):1123–30.
9. Schabereiter-Gurtner C, et al. Novel real-time PCR assay for detection of *Helicobacter pylori* infection and simultaneous clarithromycin susceptibility testing of stool and biopsy specimens. J Clin Microbiol. 2004;42(10):4512–8.
10. Agudo S, et al. Detection of *Helicobacter pylori* and clarithromycin resistance in gastric biopsies of pediatric patients by using a commercially available real-time polymerase chain reaction after NucliSens semiautomated DNA extraction. Diagn Microbiol Infect Dis. 2010;67(3):213–9.
11. Vécsei A, et al. Stool polymerase chain reaction for *Helicobacter pylori* detection and clarithromycin susceptibility testing in children. Clin Gastroenterol Hepatol. 2010;8(3):309–12.
12. Shin SY, et al. Evaluation of the Seeplex® Meningitis ACE Detection kit for the detection of 12 common bacterial and viral pathogens of acute meningitis. Ann Lab Med. 2012;32(1):44–9.

13. Hultén K, et al. Macrolide resistance in *Helicobacter pylori*: mechanism and stability in strains from clarithromycin-treated patients. Antimicrob Agents Chemother. 1997;41(11):2550–3.
14. Mégraud F. *H. pylori* antibiotic resistance: prevalence, importance, and advances in testing. Gut. 2004;53(9):1374–84.
15. Alarcón T, et al. Clarithromycin resistance stability in *Helicobacter pylori*: influence of the MIC and type of mutation in the 23S rRNA. J Antimicrob Chemother. 2000;46(4):613–6.
16. van Doorn LJ, et al. Accurate prediction of macrolide resistance in *Helicobacter pylori* by a PCR line probe assay for detection of mutations in the 23S rRNA gene: multicenter validation study. Antimicrob Agents Chemother. 2001;45(5):1500–4.
17. Maeda S, et al. Detection of clarithromycin-resistant *helicobacter pylori* strains by a preferential homoduplex formation assay. J Clin Microbiol. 2000;38(1):210–4.
18. Kato S, et al. Antibiotic resistance of *Helicobacter pylori* strains in Japanese children. J Clin Microbiol. 2002;40(2):649–53.
19. Pan ZJ, et al. Assessment of clarithromycin-resistant *Helicobacter pylori* among patients in Shanghai and Guangzhou, China, by primer-mismatch PCR. J Clin Microbiol. 2002;40(1):259–61.
20. An B, et al. Antibiotic resistance in *Helicobacter pylori* strains and its effect on *H. pylori* eradication rates in a single center in Korea. Ann Lab Med. 2013;33(6):415–9.
21. Impact of clarithromycin resistance on the outcome of standard triple *Helicobacter pylori* eradication therapy. Korean J Helicobacter Up Gastrointest Res. 2010;10(1):14–20.
22. Chung WC, et al. Dual-priming oligonucleotide-based multiplex PCR using tissue samples in rapid urease test in the detection of *Helicobacter pylori* infection. World J Gastroenterol. 2014;20(21):6547–53.
23. Ramírez-Lázaro MJ, et al. Real-time PCR improves *Helicobacter pylori* detection in patients with peptic ulcer bleeding. PLoS One. 2011;6(5):e20009.
24. Sabbagh P, et al. Diagnostic methods for *Helicobacter pylori* infection: ideals, options, and limitations. Eur J Clin Microbiol Infect Dis. 2019;38(1):55–66.
25. Patel SK, et al. *Helicobacter pylori* is not eradicated after triple therapy: a nested PCR based study. Biomed Res Int. 2014;2014:483136.
26. Ferguson DA Jr, et al. Evaluation of two string tests for obtaining gastric juice for culture, nested-PCR detection, and combined single- and double-stranded conformational polymorphism discrimination of *Helicobacter pylori*. Dig Dis Sci. 1999;44(10):2056–62.
27. Singh V, et al. Evaluation of nested PCR in detection of *Helicobacter pylori* targeting a highly conserved gene: HSP60. Helicobacter. 2008;13(1):30–4.
28. Taborda MI, et al. Non-invasive diagnostic of *Helicobacter pylori* in stools by nested-qPCR. Pol J Microbiol. 2018;67(1):11–8.
29. Li Y, et al. Evaluation of multiplex ARMS-PCR for detection of *Helicobacter pylori* mutations conferring resistance to clarithromycin and levofloxacin. Gut Pathog. 2020;12:35.
30. Wang Y, et al. Multiplex accelerated PCR system for one-step *Helicobacter pylori* cagA + genotypes detection: a guide for clinical testing. Curr Microbiol. 2022;79(8):235.

Part III

Treatment of *Helicobacter pylori*

13 Empirical Therapy

Hyun Ho Choi and Sung Soo Kim

Highlights

What Is Known

- **Standard triple therapy**
  - PPI (standard dose, bid) + amoxicillin (1 g, bid) + clarithromycin (500 mg, bid) for 14 days.
  - Application: Recommended as first-line treatment in areas with low clarithromycin resistance.
  - Effectiveness: Its efficacy is compromised by clarithromycin resistance, leading recent guidelines to recommend for a 14-day regimen.
- **Sequential therapy**
  - PPI (standard dose, bid) + amoxicillin (1 g, bid) for first 5 days, and PPI (standard dose, bid) + clarithromycin (500 mg, bid) + metronidazole (500 mg, bid) for second 5 days.
  - Application: Provided an alternative to standard triple therapy, aiming to improve eradication rates by overcoming antibiotic resistance.
  - Effectiveness: Developed to improve eradication rates through bacterial load reduction and resistance mechanism prevention, particularly against clarithromycin, showing higher efficacy than standard triple therapy.
- **Concomitant therapy**
  - PPI (standard dose, bid) + amoxicillin (1 g, bid) + clarithromycin (500 mg, bid) + metronidazole (500 mg, bid) for 10 days.
  - Application: Part of the non-bismuth quadruple therapy, intended to address the limitations of standard triple therapy by including metronidazole to combat clarithromycin resistance. It is recommended where clarithromycin resistance is high or when bismuth is unavailable.
  - Effectiveness: Achieves higher eradication rates compared to standard triple therapy, making it a potent option against *H. pylori* infections.

Future Research Questions

- More effective regimen is anticipated.
- Optimal duration should be decided.
- New antibiotics should be developed.

H. H. Choi · S. S. Kim (✉)
Department of Internal Medicine, Uijeongbu St. Mary's Hospital, The Catholic University of Korea, Seoul, Korea
e-mail: kimss@catholic.ac.kr

B.-W. Kim (ed.), *Helicobacter Infection in Clinical Practice*,
https://doi.org/10.1007/978-981-97-8397-7_13

13.1 Introduction

To qualify as a first-line treatment for *Helicobacter pylori* eradication therapy, a regimen should demonstrate an eradication rate exceeding 85% in an intention-to-treat (ITT) analysis and surpassing 90% in a per-protocol (PP) analysis [1, 2]. However, numerous studies indicate that the eradication rate of standard triple therapy falls below 80%. Consequently, alternative regiments, incorporating diverse drug combinations varied durations of drug administration, have been explored to counter the declining efficacy of standard triple therapy.

For individuals undergoing *H. pylori* eradication for the first time, four potential regimens are considered: 14-day standard triple therapy, non-bismuth quadruple therapy (sequential therapy, concomitant therapy), 7-day standard triple therapy following clarithromycin resistance testing, and bismuth quadruple therapy [1, 3–6]. To appropriately choose *H. pylori* eradication therapy, one should take into account antibiotic resistance and the eradication rates associated with the specific drug treatment.

In this chapter, we will focus on empirical treatments for *H. pylori*, emphasizing standard triple therapy, sequential therapy, and concomitant therapy. Further details on bismuth quadruple therapy will be presented in a subsequent chapter.

13.2 Main Body

13.2.1 Standard Triple Therapy

The most commonly employed first-line *H. pylori* eradication therapy is standard triple therapy, comprising a standard dose proton-pump inhibitor (PPI), amoxicillin (1 g), and clarithromycin (500 mg) administered twice daily. However, lots of reports indicate an unsatisfactory success rate for the current standard triple therapy. Factors contributing to eradication failure include high bacterial load, elevated intragastric acidity, the virulence of *H. pylori* strains, and poor treatment compliance. Clarithromycin is the pivotal antibiotic in the standard triple regimen, and numerous studies highlight its resistance as a significant factor in the failure of standard triple treatment [7, 8].

Recently, primary clarithromycin resistance in *H. pylori* has emerged as a significant concern globally [9]. Clarithromycin resistance rates are on the rise worldwide, with susceptibility to clarithromycin varying across regions. In a systematic review and meta-analysis by Savoldi et al. [9], incorporating 178 studies from 65 countries with a total of 66,142 isolates, primary clarithromycin resistance exceeded 15% in the European region (18%; 95% CI, 16–20%), the Eastern Mediterranean Region (33%; 95% CI, 23–44%), and the Western Pacific region (34%; 95% CI, 30–38%). In contrast, the Americas Region (10%: 95% CI, 4–16%) and the South-East Asia Region (10%: 95% CI, 5–16%) exhibited lower prevalence of primary clarithromycin resistance.

In a multicenter nationwide antibiotic resistance study in Korea, Lee et al. reported a clarithromycin resistance rate of 17.8%, with a multidrug resistance rate for *H. pylori* eradication drugs reaching 25.2% [10]. Additionally, Megraud et al. found an overall clarithromycin resistance rate of 17.5%, with higher resistance rates in Western/Central/Southern Europe (>20%) compared to Northern European countries (<10%) [11].

While several guidelines endorse the use of standard triple therapy as the preferred first-line treatment, variations exist in the perceptions of this approach among guidelines in different regions. Specifically, the Maastricht VI/Florence consensus guidelines recommend employing standard triple therapy as the first-line empirical treatment when clarithromycin resistance is less than 15%, with a recommended treatment duration of 14 days [4]. In North America, standard triple therapy is recommended as a first-line treatment option for *H. pylori*, especially in regions where clarithromycin resistance is below 15%, and for individuals with no prior macrolide exposure or proven high local eradication rates (>85%) [1, 3]. The Japanese guidelines advocate standard triple therapy for 7 days, recommending *H. pylori* testing for clarithromycin resistance

[5]. If a susceptibility test is not performed, clarithromycin may be substituted with metronidazole. In Korea, a 14-day standard triple therapy is recommended for first-line regimens in the absence of clarithromycin resistance testing [6].

Guidelines recommend a 14-day duration for standard triple therapy [1, 3, 4, 6]. In the Maastricht VI/Florence consensus guidelines, the European Registry on *Helicobacter pylori* management (Hp-EuReg) reported that a 14-day therapy remains effective until clarithromycin resistance exceeds approximately 15%, reporting eradication rates in the modified intention-to-treat as follows: 7-day (82%), 10-day (83%), and 14-day (87%) [4]. The Toronto Consensus and American College of Gastroenterology (ACG) guidelines, as analyzed by Yuhong et al. in a meta-analysis, demonstrated a higher eradication rate with a 14-day regimen compared to 7-day and 10-day regimens. For the 14 vs. 7 days comparison (34 studies, Relative risk (RR) 0.65, 95% CI 0.57–0.75; number needed to treat (NNT) 12, 95% CI 9–16), and the 14 vs. 10 days comparison (10 studies, RR 0.69, 95% CI 0.52–0.91) [12].

According to the Korean guideline [6], the pooled eradication rate for the 14-day therapy was 78.1% (95% CI, 75.2–80.7%), significantly higher than the rates for 7-day (70.0%, 95% CI, 68.5–71.4%) and 10-day therapy (73.7%, 95% CI, 69.8–77.2%) ($P < 0.01$ for both duration) in an intention to treat (ITT) analysis. Eradication rates between the 7-day and 10-day therapy were not significantly different. Additionally, Jung et al. [13], in their network meta-analysis, reported that the ITT analysis showed eradication rates of 7-day, 10-day, and 14-day standard triple therapy as 71.1% (95% CI, 68.3–73.7%), 67.0% (95% CI, 60.0–73.4%), and 76.4% (95% CI, 73.3–79.2%), respectively.

13.2.2 Sequential Therapy

Sequential therapy involves administering a PPI and amoxicillin for the first 5 days, followed by a PPI and two other antibiotics for the subsequent 5 days. Originally introduced by Zullo et al. in 2000 [14], this 10-day sequential treatment demonstrated a high eradication rate of 98% based on ITT analysis. The rationale behind sequential therapy lies in the observation that a superior eradication rate could be achieved by utilizing a standard triple-drug regimen for 7 days in patients who had previously failed eradication following 14 days of PPI and amoxicillin two-drug therapy [15]. Although the precise mechanism of the sequential therapy is unknown, two reasons are proposed to explain its anticipated high eradication rate [16]. Firstly, it is well known that the efficacy of eradication treatment is inversely proportional to the density of *H. pylori* in the stomach. In sequential therapy, the administration of PPI and amoxicillin in the initial phase achieves a cure rate of up to 50%. The efficacy of a triple therapy is inversely correlated with bacterial load, demonstrating higher eradication rates in individuals with low bacterial density in the stomach. Secondly, one of the primary mechanisms of clarithromycin resistance in *H. pylori* eradication treatment is the development of efflux channels by *H. pylori*, which rapidly expel the drug from the bacterial cell, hindering the binding of the antibiotic to the ribosome. Amoxicillin, administered in the first half of sequential therapy, acts on the bacterial cell wall and induces damage. This action may prevent the development of efflux channels by weakening the cell wall of *H. pylori*.

The eradication rate of 10-day sequential therapy as a first-line treatment was higher than that of the 7-day and 10-day standard triple therapy. However, there was no significant difference in eradication rates between sequential therapy and 14 days of standard triple therapy or bismuth quadruple therapy. In a network meta-analysis by Gatta et al. [17], the eradication rate of sequential therapy was 84.3% (95% confidence interval 82.1–86.4%). Notably, sequential therapy achieved significantly higher eradication rates compared to 7 or 10 days standard triple therapy (relative risk 1.21, 95% confidence interval 1.17–1.25; 1.11, 1.04–1.19, respectively). However, sequential therapy did not demonstrate superior efficacy compared to 14 days of standard triple therapy, concomitant therapy, or bismuth quadruple therapy (relative risk 1.00,

0.94–1.06; 0.99, 0.94–1.05; 1.01, 0.95–1.06, respectively). According to Korean guidelines [6], the eradication rates of sequential therapy were 84.1% in ITT analysis and 85.9% in PP analysis. In contrast, the overall standard triple therapy achieved eradication rates of 74.9% in ITT analysis and 77.3% in PP analysis. The pooled RR of eradication rates (sequential therapy vs standard triple therapy) was 1.37 (95% CI, 1.21–1.54) in ITT and 1.60 (95% CI, 1.40–1.83) in PP using the fixed effects model, respectively.

Various guidelines offer distinct recommendations for sequential therapy due to regional factors, drug resistance patterns, and variations in available evidence. Clarithromycin resistance, metronidazole resistance, and the combination of clarithromycin and metronidazole resistance all diminish the effectiveness of sequential therapy. Consequently, some guidelines do not recommend sequential therapy as the first-line treatment [1, 4, 18, 19]. ACG guidelines acknowledge that while it may not be the ideal choice, it is still deemed acceptable for clinical practice and categorizes sequential therapy as "suggested" rather than "recommended" [3]. On the other hand, Korean guidelines recommend sequential therapy as a first-line treatment when a clarithromycin resistance test is not available [6].

The characteristics of sequential treatment is that the medications taken in the first half and the second half are different, necessitating a change in medication halfway through the treatment course. Due to the complexity of the sequential therapy regimen, patients may encounter confusion in taking their medications, potentially affecting their compliance. This factor should be taken into consideration when implementing sequential therapy in actual clinical practice.

13.2.3 Concomitant Therapy

Concomitant therapy involves the simultaneous administration of clarithromycin, amoxicillin, metronidazole, and a PPI. This treatment strategy employs three antibiotics within a short time frame to reduce cost, to minimize adverse events, and to mitigate antibiotic resistance. The Maastricht VI/Florence consensus report recommends concomitant therapy in cases where bismuth is unavailable, the rate of clarithromycin resistance is known to be high (>15%), or susceptibility testing is not available [15]. Various guidelines, including those from the ACG, the Toronto Consensus, and the Korean guidelines, endorse concomitant therapy as an optional choice for the first-line treatment of *H. pylori* infection [1, 3, 6]. When utilizing concomitant therapy for *H. pylori*, the recommended duration is 10–14 days. In a meta-analysis of 19 randomized trials encompassing 2070 patients, Gisbert et al. [20] reported significantly higher eradication rates with concomitant therapy (90%) compared to standard triple therapy (78%) in ITT analysis (OR, 2.36; 95% CI, 1.67–3.34). A trend toward higher cure rates was observed with longer durations of therapy (7–10 days versus 3–5 days), leading to a recommendation for the optimal treatment length of 10 days. According to the Korean guidelines [6], the eradication rates of concomitant therapy for 10-day and 14-day regimen were 85% and 86% in ITT analysis, and 91% and 94% in PP analysis, respectively. These results outperformed those of standard triple therapy. In subgroup analysis, there was no significant difference in eradication rates according to the duration (10-day, $P = 0.925$ in ITT analysis; 14-day, $P = 0.278$).

The efficacy of concomitant therapy was decreased in patients with clarithromycin-resistant *H. pylori* infection, albeit to a lesser extent when compared to standard triple therapy [20]. Numerous reports indicate that standard triple therapy exhibits significantly reduced efficacy in regions with high clarithromycin resistance, with concomitant therapy surpassing sequential therapy in effectiveness [1, 4, 18, 21]. In a recent randomized study conducted in Korea, comparing the eradication rates of concomitant treatment and sequential therapy, the eradication rate of concomitant therapy was significantly higher at 93.3% vs. 53.9% ($P = 0.029$) when specifically analyzing clarithromycin-resistant strains [22].

The optimum duration for concomitant therapy still remains a topic of debate. The Maastricht consensus recommends 14-day treatment, unless locally effective 10-day therapies are established. In Greece, Apostolopoulos et al. [23] reported overall eradication rates for 10-day and 14-day concomitant therapy as 87.9% vs. 87.4% in the ITT analysis ($P = 1.000$), and 93% vs. 94.1% in the PP analysis ($P = 0.859$), respectively, with similar compliance. Romano et al. [24] demonstrated significantly higher eradication rates with 14-day therapy compared to 10-day therapy in both ITT (96.1% vs 80%, $P = 0.001$) and PP (97% vs 85.1%, $P = 0.003$) analysis. Nyssen et al. [25], in the Hp-EuReg study involving 4164 patients prescribed with concomitant therapy, reported modified ITT eradication rates of 88.3% for the 10-day regimen and 92.1% for the 14-day regimen.

Concomitant therapy is less complicated compared to sequential therapy, but some reports highlight a higher frequency of adverse events associated with concomitant therapy. Additionally, if *H. pylori* eradication is unsuccessful with concomitant therapy, selecting a secondary eradication regimen becomes a challenging task.

13.3 Conclusion

The efficacy of *H. pylori* eradication increasingly relies on personalized treatment strategies that counteract rising antibiotic resistance, especially to clarithromycin, a major cause to treatment failure. Therefore, it is important to choose the eradication regimen based on eradication rates, local antibiotic resistance patterns, the specific characteristics of each therapy, and individual patient factors. Standard triple therapy is recommended as the first-line treatment in areas with low clarithromycin resistance. However, with the effectiveness of standard triple therapy waning globally due to this resistance, alternative therapies such as sequential and concomitant have emerged. Sequential therapy offers a dual-phase treatment strategy that potentially enhances eradication rates by reducing bacterial load and preventing resistance mechanisms. Concomitant therapy, proposing a simultaneous administration of multiple antibiotics, aims to improve success rates and reduce the impact of clarithromycin resistance. Comparison of each regimen is summarized in Table 13.1. A comprehensive understanding of the effectiveness and characteristics of each therapy is essential for choosing the most appropriate treatment.

Table 13.1 Regimen of empirical therapy for *Helicobacter pylori* infection

Regimen	Contents
Standard triple therapy	PPI (standard dose, bid) + amoxicillin (1 g, bid) + clarithromycin (500 mg, bid) for 14 days
Sequential therapy	PPI (standard dose, bid) + amoxicillin (1 g, bid) for first 5 days, and PPI (standard dose, bid) + clarithromycin (500 mg, bid) + metronidazole (500 mg, bid) for second 5 days
Concomitant therapy	PPI (standard dose, bid) + amoxicillin (1 g, bid) + clarithromycin (500 mg, bid) + metronidazole (500 mg, bid) for 10 days

PPI proton-pump inhibitor, *bid* twice per day

References

1. Fallone CA, Chiba N, van Zanten SV, et al. The Toronto Consensus for the treatment of Helicobacter pylori infection in adults. Gastroenterology. 2016;151:51–69 e14.
2. Graham DY, Lu H, Yamaoka Y. A report card to grade Helicobacter pylori therapy. Helicobacter. 2007;12:275–8.
3. Chey WD, Leontiadis GI, Howden CW, Moss SF. ACG clinical guideline: treatment of Helicobacter pylori infection. Am J Gastroenterol. 2017;112:212–39.
4. Malfertheiner P, Megraud F, Rokkas T, et al. Management of Helicobacter pylori infection: the Maastricht VI/Florence consensus report. Gut. 2022;
5. Kato M, Ota H, Okuda M, et al. Guidelines for the management of Helicobacter pylori infection in Japan: 2016 Revised Edition. Helicobacter. 2019;24:e12597.
6. Jung HK, Kang SJ, Lee YC, et al. Evidence-based guidelines for the treatment of Helicobacter pylori infection in Korea 2020. Gut Liver. 2021;15:168–95.
7. Nyssen OP, McNicholl AG, Megraud F, et al. Sequential versus standard triple first-line therapy for Helicobacter pylori eradication. Cochrane Database Syst Rev. 2016;2016:Cd009034.
8. Braden B. The best and worst treatments for Helicobacter pylori. BMJ. 2015;351:h4146.
9. Savoldi A, Carrara E, Graham DY, Conti M, Tacconelli E. Prevalence of antibiotic resistance in Helicobacter pylori: a systematic review and meta-analysis in World Health Organization regions. Gastroenterology. 2018;155:1372–1382 e1317.
10. Lee JH, Ahn JY, Choi KD, et al. Nationwide antibiotic resistance mapping of Helicobacter pylori in Korea: a prospective multicenter study. Helicobacter. 2019;24:e12592.
11. Megraud F, Coenen S, Versporten A, et al. Helicobacter pylori resistance to antibiotics in Europe and its relationship to antibiotic consumption. Gut. 2013;62:34–42.
12. Yuan Y, Ford AC, Khan KJ, et al. Optimum duration of regimens for Helicobacter pylori eradication. Cochrane Database Syst Rev. 2013:CD008337.
13. Jung YS, Park CH, Park JH, Nam E, Lee HL. Efficacy of Helicobacter pylori eradication therapies in Korea: a systematic review and network meta-analysis. Helicobacter. 2017:22.
14. Zullo A, Rinaldi V, Winn S, et al. A new highly effective short-term therapy schedule for Helicobacter pylori eradication. Aliment Pharmacol Ther. 2000;14:715–8.
15. Rinaldi V, Zullo A, Pugliano F, Valente C, Diana F, Attili AF. The management of failed dual or triple therapy for Helicobacter pylori eradication. Aliment Pharmacol Ther. 1997;11:929–33.
16. Vaira D, Zullo A, Hassan C, Fiorini G, Vakil N. Sequential therapy for Helicobacter pylori eradication: the time is now! Therap Adv Gastroenterol. 2009;2:317–22.
17. Gatta L, Vakil N, Vaira D, Scarpignato C. Global eradication rates for Helicobacter pylori infection: systematic review and meta-analysis of sequential therapy. BMJ. 2013;347:f4587.
18. Liu WZ, Xie Y, Lu H, et al. Fifth Chinese National Consensus Report on the management of Helicobacter pylori infection. Helicobacter. 2018;23:e12475.
19. Fallone CA, Moss SF, Malfertheiner P. Reconciliation of recent Helicobacter pylori treatment guidelines in a time of increasing resistance to antibiotics. Gastroenterology. 2019;157:44–53.
20. Gisbert JP, Calvet X. Update on non-bismuth quadruple (concomitant) therapy for eradication of Helicobacter pylori. Clin Exp Gastroenterol. 2012;5:23–34.
21. Graham DY, Dore MP. Helicobacter pylori therapy: a paradigm shift. Expert Rev Anti Infect Ther. 2016;14:577–85.
22. Kim SY, Lee SW, Choe JW, et al. Helicobacter pylori eradication rates of concomitant and sequential therapies in Korea. Helicobacter. 2017;22
23. Apostolopoulos P, Ekmektzoglou K, Georgopoulos S, et al. 10-Day versus 14-day quadruple concomitant nonbismuth therapy for the treatment of Helicobacter pylori infection: results from a randomized prospective study in a high clarithromycin resistance country. J Clin Gastroenterol. 2020;54:522–7.
24. Romano M, Gravina AG, Nardone G, et al. Non-bismuth and bismuth quadruple therapies based on previous clarithromycin exposure are as effective and safe in an area of high clarithromycin resistance: a real-life study. Helicobacter. 2020;25:e12694.
25. Nyssen OP, Bordin D, Tepes B, et al. European Registry on Helicobacter pylori management (Hp-EuReg): patterns and trends in first-line empirical eradication prescription and outcomes of 5 years and 21 533 patients. Gut. 2021;70:40–54.

14 Tailored Therapy

Byung-Wook Kim

Highlights

What Is Known
• Culture-based therapy forms the fundamental approach for treating bacterial infection. • PCR-based tailored therapy has demonstrated superior efficacy compared to empirical therapy.
Future Research Questions
• The optimal duration for tailored therapy of *H. pylori* remains undetermined. • The diagnostic accuracy and cost-effectiveness of each method should be elucidated.

14.1 Introduction

The eradication rate of empirical triple therapy for *Helicobacter pylori* has been declining over the decades due to the increasing prevalence of clarithromycin resistance [1]. Precise prediction and diagnosis of the clarithromycin-resistance strain are crucial for the successful primary eradication of *H. pylori*. To address this challenge, numerous new therapeutic strategies have been emerged. The optimal treatment approach involves combining antibiotics based on susceptibility test results. However, the efficacy of *H. pylori* culture is often suboptimal, leading to the reliance on point mutation detection, a closely linked marker of resistance. In this chapter, I aim to delve into the topic of tailored therapy focused on clarithromycin resistance.

B.-W. Kim (✉)
Department of Internal Medicine, Incheon St. Mary's Hospital, The Catholic University of Korea, Seoul, Korea
e-mail: gastro@catholic.ac.kr

14.2 Main Body

14.2.1 Culture-Based Tailored Therapy

Culture-based therapy forms the fundamental approach for treating bacterial infection. The measurement of minimal inhibitory concentration (MIC) via culturing gastric biopsy samples boasts a rich history as an antibiotic sensitivity test [2]. Nevertheless, the success rate of the *H. pylori* culture is suboptimal, and obtaining the result of *H. pylori* culture along with antibiotic sensitivity test takes at least 1 week. Methods for determining MIC encompass the agar dilution method, broth microdilution method, and the E-test. While the agar dilution method is time consuming, it is regarded as the gold standard for antibiotic sensitivity testing [3]. The E-test, on the other hand, is straightforward and widely employed in clinical practice [4].

B.-W. Kim (ed.), *Helicobacter Infection in Clinical Practice*,
https://doi.org/10.1007/978-981-97-8397-7_14

A meta-analysis comparing susceptibility-guided treatment to empirical treatment demonstrated that culture-based therapy yielded better outcomes in both first-line and second-line treatment [5]. In a Korean study, it was observed that 14-day tailored therapy based on *H. pylori* culture as a second-line eradication regimen produced superior results compared to a 14-day empirical therapy [6]. Nevertheless, culture-based therapy is not feasible in primary care clinics, severely limiting its clinical utility.

14.2.2 PCR-Based Tailored Therapy

Clarithromycin exerts its antibacterial activity by binding to the peptidyl transferase region of *H. pylori*'s 23S rRNA [7]. Point mutations in the *rrl* gene, responsible for encoding the binding region of 23S rRNA (particularly A2142G, A2143G, or A2144G point mutation), contribute to clarithromycin resistance [8, 9]. These mutations can be identified through PCR and other detection techniques. PCR methods for detecting these point mutations include real-time PCR, multiplex PCR, and nested PCR. The diagnostic accuracy of PCR methods is on par with agar dilution methods [10, 11].

Current guidelines for the treatment of *H. pylori* advocate tailored therapy to improve eradication rate. Several studies have compared tailored therapy to empirical therapy [12–14], and PCR-based tailored therapy has demonstrated superior efficacy compared to empirical therapy. It is worth noting that the empirical therapy regimens varied significantly in these studies. A retrospective study demonstrated that 14-day empirical treatment outperformed a 7-day therapy, but there was no discernible distinction between a 14-day and a 7-day regimen in PCR-based tailored therapy [15]. Nonetheless, the optimal duration for tailored therapy remains undetermined, and further research is anticipated to elucidate the most effective tailored therapy regimen.

14.2.3 Next-Generation Sequencing

Next-generation sequencing (NGS) offers a comprehensive assessment of mutations related to specific antimicrobial agents, including genes associated with resistance to clarithromycin, amoxicillin, metronidazole, and levofloxacin [16–19]. Moreover, it enables the concurrent detection of mutations linked to multiple antimicrobial resistances. NGS also effectively evaluates resistance to rifabutin and tetracycline. However, its findings show limited concordance with culture-based methods in the cases of amoxicillin and metronidazole resistance [20]. It is important to note that a drawback of NGS is its higher cost compared to other testing methods.

14.2.4 Cost-Effectiveness of Tailored Therapy

Tailored therapy necessitates additional procedures like culture or PCR testing, leading to increased costs. Previous studies comparing tailored therapy and empirical therapy typically computed costs for individual instances of successful eradication or the total cost for all procedures [21, 22]. Nevertheless, factors such as shortened treatment duration, mitigation of the propagation of antibiotic-resistant strains, and the post-tailored therapy financial implications should also should be taken into account.

14.3 Conclusions

Considering that culture or PCR-based antibiotic resistance tests are not available in every clinic, empirical therapy is inevitable. Nevertheless, susceptibility-based treatment demonstrated better outcome compared to empirical treatment. To improve the eradication rate, every effort should be done to obtain a susceptibility-based treatment. The advantages and disadvantages of each susceptibility test is described in Table 14.1.

Table 14.1 Advantages and disadvantages of susceptibility tests

	Advantages	Disadvantages
Culture-based test	Can test every antibiotic resistance	Low culture success rate Time consuming Expensive
PCR-based test	Relatively low cost Rapid result Equivocal efficacy compared to culture-based test Affordable in PPI users	Not all antibiotics can be tested False positive from dead bacteria
NGS-based test	Enabled the detection of novel, rare and complex resistance mechanisms	Highly expensive

PCR polymerase chain reaction, *NGS* next-generation sequencing, *PPI* proton-pump inhibitor

References

1. Kim YK, Kim JS, Kim BW. Recent trends of *Helicobacter pylori* eradication therapy in Korea. Korean J Helicobacter Up Gastrointest Res. 2012;12:219–23.
2. Osato MS. Antimicrobial susceptibility testing for *Helicobacter pylori*: sensitivity test results and their clinical relevance. Curr Pharm Des. 2000;6:1545–55.
3. Valdivieso-García A, Imgrund R, Deckert A, et al. Cost analysis and antimicrobial susceptibility testing comparing the E test and the agar dilution method in *Campylobacter jejuni* and *Campylobacter coli*. Diagn Microbiol Infect Dis. 2009;65:168–74.
4. Osato MS, Reddy R, Reddy SG, Penland RL, Malaty HM, Graham DY. Pattern of primary resistance of *Helicobacter pylori* to metronidazole or clarithromycin in the United States. Arch Intern Med. 2001;161:1217–20.
5. López-Góngora S, Puig I, Calvet X, et al. Systematic review and meta-analysis: susceptibility-guided versus empirical antibiotic treatment for *Helicobacter pylori* infection. J Antimicrob Chemother. 2015;70:2447–55.
6. Kwon YH, Kim N, Lee JY, et al. Comparison of the efficacy of culture-based tailored therapy for *Helicobacter pylori* eradication with that of the traditional second-line rescue therapy in Korean patients: a prospective single tertiary center study. Scand J Gastroenterol. 2016;51:270–6.
7. Arslan N, Yılmaz Ö, Demiray-Gürbüz E. Importance of antimicrobial susceptibility testing for the management of eradication in *Helicobacter pylori* infection. World J Gastroenterol. 2017;23:2854–69.
8. Mégraud F. Resistance of *Helicobacter pylori* to antibiotics. Aliment Pharmacol Ther. 1997;11(Suppl 1):43–53.
9. Zanotti G, Cendron L. Structural aspects of *Helicobacter pylori* antibiotic resistance. Adv Exp Med Biol. 2019;1149:227–41.
10. Noh JH, Ahn JY, Choi J, et al. Real-time polymerase chain reaction for the detection of *Helicobacter pylori* and clarithromycin resistance. Gut Liver. 2023;17:75–381.
11. Saruuljavkhlan B, Yamaoka Y. Benefits of a molecular-based method for the detection of clarithromycin-resistant *Helicobacter pylori*. Gut Liver. 2021;15:487–9.
12. Kawai T, Yamagishi T, Yagi K, et al. Tailored eradication therapy based on fecal *Helicobacter pylori* clarithromycin sensitivities. J Gastroenterol Hepatol. 2008;23(Suppl 2):S171–4.
13. Kim JL, Cho SJ, Chung SJ, et al. Empiric versus clarithromycin resistance-guided therapy for *Helicobacter pylori* based on polymerase chain reaction results in patients with gastric neoplasms or gastric mucosa-associated lymphoid tissue lymphoma: a randomized controlled trial. Clin Transl Gastroenterol. 2020;11:e00194.
14. Kim SJ, Jee SR, Park MI, et al. A randomized controlled trial to compare *Helicobacter pylori* eradication rates between the empirical concomitant therapy and tailored therapy based on 23S rRNA point mutations. Medicine (Baltimore). 2022;101:e30069.
15. Na SY, Kim BW, Kim MJ, Choe Y, Kim JS. Effective eradication regimen and duration according to the clarithromycin susceptibility of *Helicobacter pylori* determined using dual priming oligonucleotide-based multiplex polymerase chain reaction. Gut Liver. 2023;17:722–30.
16. Chen J, Ye L, Jin L, et al. Application of next-generation sequencing to characterize novel mutations in clarithromycin-susceptible *Helicobacter pylori* strains with A2143G of 23S rRNA gene. Ann Clin Microbiol Antimicrob. 2018;17:10.
17. Qureshi NN, Gallaher B, Schiller NL. Evolution of amoxicillin resistance of *Helicobacter pylori* in vitro: characterization of resistance mechanisms. Microb Drug Resist. 2014;20:509–16.
18. Binh TT, Suzuki R, Trang TT, Kwon DH, Yamaoka Y. Search for novel candidate mutations for metronidazole resistance in *Helicobacter pylori* using next-generation sequencing. Antimicrob Agents Chemother. 2015;59:2343–8.

19. Miftahussurur M, Shrestha PK, Subsomwong P, Sharma RP, Yamaoka Y. Emerging *Helicobacter pylori* levofloxacin resistance and novel genetic mutation in Nepal. BMC Microbiol. 2016;16:256.
20. Hulten KG, Genta RM, Kalfus IN, Zhou Y, Zhang H, Graham DY. Comparison of culture with antibiogram to next-generation sequencing using bacterial isolates and formalin-fixed, paraffin-embedded gastric biopsies. Gastroenterology. 2021;161:1433–42.
21. Choe AR, Shim KN, Park Y, et al. Cost-effectiveness, efficacy, and safety analysis of tailored therapy in patients with *Helicobacter pylori* infection. J Clin Med. 2021;10:2619.
22. Cho JH, Jin SY, Park S. Comparison of tailored *Helicobacter pylori* eradication versus modified bismuth quadruple therapy in Korea: a randomized controlled trial. Expert Rev Anti Infect Ther. 2022;20:923–9.

15 Proton-Pump Inhibitors and Potassium-Competitive Acid Blockers

Chan Hyuk Park

Highlights

What Is Known
• Proton-pump inhibitors (PPIs) are effectively used in *H. pylori* eradication therapy, demonstrating high treatment success rates even in cases of resistant infections. • Gastric acid suppressants enhance the efficacy of antimicrobials within the stomach and increase susceptibility to antibiotics by elevating gastric acidity levels. • Potassium-competitive acid blockers(P-CABs), especially vonoprazan, present notable advantages over PPIs in the treatment *H. pylori* eradication, caused by their faster onset of action and long-lasting acid-inhibitory potency.
Future Research Questions
• We need to further evaluate the efficacy of emerging P-CABs like tegoprazan to understand its potential for surpassing PPI efficacy in *H. pylori* eradication. • Additionally, investigating the optimal dosage of tegoprazan for enhanced efficacy in treating *H. pylori* infection may be required.

15.1 Introduction

Approximately three decades ago, proton-pump inhibitors (PPIs) became accessible and started being utilized to eliminate *Helicobacter pylori* infection [1]. Eradication treatments that include PPIs have shown higher effectiveness compared to those that do not include them [2], and eradication therapy regimens, which consist of PPI and two or more antibiotics, are currently used worldwide [3, 4]. In this section, we will review the role and efficacy of gastric acid suppressants, including PPIs and potassium-competitive acid blockers (P-CABs), in the treatment of *H. pylori* infection.

C. H. Park (✉)
Department of Internal Medicine, Hanyang University Guri Hospital, Hanyang University College of Medicine, Guri, Korea

15.2 Main Body

Gastric acid suppressants, including PPIs, have several roles in the treatment of *H. pylori* infection [5]: improving antibiotic efficacy by stabilizing acid-labile antibiotics, increasing the concentration of antibiotics in the gastric juice, and enhancing the sensitivity of *H. pylori* to antibiotics. The association between intra-gastric acidity and the successful eradication of *H. pylori* is well known. In a study on the dual therapy for *H. pylori* infection, 24% of the ongoing-infection group and 6% of the cured-infection group had an intra-gastric pH under 4, respectively [6]; thus, the intra-gastric pH was higher in the cured-infection group than that in the ongoing-infection group. A study on triple therapy also showed that intra-gastric pH was higher in the successful-eradication group than in the failed-eradication group [7], which had a median

B.-W. Kim (ed.), *Helicobacter Infection in Clinical Practice*,
https://doi.org/10.1007/978-981-97-8397-7_15

intra-gastric pH < 4.5. The treatment has failed in patients who had been infected with clarithromycin-resistant *H. pylori* even though their intra-gastric pH reached approximately 6; interestingly, some patients with a median intra-gastric pH > 6.5 were successfully treated despite clarithromycin-resistant *H. pylori* infection. These results imply that strong acid inhibition improves the success rate of *H. pylori* eradication therapies.

However, PPIs have limitations in terms of acid inhibition [8, 9]. First, the active forms of PPIs are unstable under acidic conditions; owing to their short half-life, PPIs require specific timing of dosing with meals, usually before breakfast. Second, approximately 25% of the proton pumps are newly biosynthesized each day. Therefore, the maximum effect of PPIs requires several days to be achieved. The inconsistent effect, which depends on the status of the cytochrome P450 2C19 of the patient, is also a concern for acid inhibition using PPIs. However, a new class of gastric acid suppressants, P-CABs, has recently emerged [10]. P-CABs do not require acid activation. They are rapidly absorbed in the small intestine, accumulate in the canalicular membranes of parietal cells [11], and inhibit proton pumps in a potassium-competitive manner [11]. As P-CABs are more stable than PPIs in the canaliculi, they allow fast and stable inhibition of gastric acid secretion. Compared to PPIs, P-CABs have several advantages in acid inhibition [10]. Generally, P-CABs have a fast onset of action, do not require specific dosing timing with food, and have less variability between patients because they are not metabolized by cytochrome P450 2C19.

P-CABs have a long development history [12–15]. The prototype of P-CABs was introduced in the 1980s [12], and all the early P-CABs were imidazopyridine derivatives. Despite its fast development [13–15], the majority of P-CABs had not been clinically used, due to hepatotoxicity, relatively low efficacy, and cost, until recently, when new formulations, including vonoprazan and tegoprazan, were approved [10, 16]. Vonoprazan, a pyrrolopyridine derivative with a long-lasting acid inhibition effect, was approved for use in Japan in 2014 [10]. In addition, tegoprazan, a benzimidazole derivative with a rapid onset of action, was approved for use in Korea in 2018 [16].

Several attempts have been made to improve the *H. pylori* eradication success rate using P-CABs instead of PPIs. Since vonoprazan was the first P-CAB approved for the treatment of *H. pylori* infection, early studies on P-CAB-based therapies were performed using vonoprazan. The first study on the efficacy of first-line *H. pylori* eradication therapy comparing P-CAB and PPI-based triple regimens was conducted in Japan and published in 2016 [17]. The study evaluated 175 patients with a 7-day vonoprazan-based triple therapy and 175 patients with a 7-day PPI-based triple therapy, which were retrospectively compared after propensity score matching. The eradication rates in the intention-to-treat analyses were 89% and 71% in the vonoprazan and PPI groups, respectively. A change in acid inhibitors alone could lead to a 20% increase in the eradication rate. Although most adverse events did not differ between groups, skin rash was more common in the vonoprazan than in the PPI group. However, the causal relationship between vonoprazan and skin rashes remains unclear. In this study, a lymphocyte transformation test was performed on 15 patients with skin rashes. None of the patients tested positive for vonoprazan or PPIs, and only eight patients tested positive for amoxicillin. A randomized controlled trial comparing 7-day vonoprazan- and 7-day lansoprazole-based triple therapies was also reported in 2016 [18]. In this study, clarithromycin-resistant *H. pylori* strains were identified in approximately 30% of the patients in both groups. Due to the high clarithromycin resistance rate, the eradication rate of lansoprazole-based triple therapy was only 76%. However, the eradication rate of the vonoprazan-based triple therapy was 93%, representing a 17% increase in the eradication rate compared to that of the lansoprazole-based triple therapy. A subgroup analysis according to clarithromycin resistance was performed to better understand why the eradication rate increased. In the subgroup of clarithromycin-susceptible strains, the eradication rates of the vonoprazan and lansoprazole groups did not differ (vonoprazan, 97.6% vs. lansoprazole, 97.3%). However, in the clarithromycin-resistant subgroup, the eradication rate in the vonoprazan group (82.0%) was more than twice that of the lansoprazole group (40.0%). Thus, the high eradication rate of the vonoprazan-based triple therapy was mainly due to the increased effect on clarithromycin-resistant strains. Regarding adverse events, diarrhea, nasopharyngitis, and dysgeusia were common in both groups, with no significant difference. A meta-analysis of ten studies on vonoprazan-based triple therapy also demonstrated that this therapy had an eradication rate 19% higher than that of the PPI-based triple therapy [19].

The efficacy of vonoprazan-based eradication therapy has also been evaluated in a recent randomized controlled trial in the US and Europe [20], in which a 14-day vonoprazan-based dual therapy, consisting of vonoprazan and amoxicillin, and a 14-day vonoprazan-based triple therapy, consisting of vonoprazan, amoxicillin, and clarithromycin, were compared with a 14-day lansoprazole-based triple therapy. Both vonoprazan-based dual and triple therapies showed superior efficacy compared to that of lansoprazole-based triple therapy (vonoprazan dual, 77.2%; vonoprazan triple, 80.8%; and lansoprazole triple therapy, 68.5%). Additionally, the primary outcome of this study was that vonoprazan-based dual and triple therapies have been proven non-inferior to lansoprazole-based triple therapy in patients with clarithromycin- and amoxicillin-susceptible strains (vonoprazan dual, 78.5%; vonoprazan triple, 84.7%; and lansoprazole triple therapy, 78.8%). The superior efficacy of vonoprazan-based therapy was significant in patients infected with clarithromycin-resistant strains. The *H. pylori* eradication rates were 69.6%, 65.8%, and 31.9% in the vonoprazan-based dual, vonoprazan-based triple, and lansoprazole-based triple therapy groups, respectively. Even in Western populations, vonoprazan is nearly twice as effective as the PPI-based triple therapy for the treatment of clarithromycin-resistant *H. pylori* infections.

Since 2021, tegoprazan, which is a new P-CAB, has been available for use in the treatment of *H. pylori* infection in Korea [21]. Although the efficacy of tegoprazan-based therapy was comparable to that of PPI-based therapy, the superior efficacy of tegoprazan over PPI-based therapy has not been identified. A multicenter randomized controlled trial showed similar eradication rates between the 7-day tegoprazan- (62.9%) and 7-day lansoprazole (60.6%)-based triple therapies. Since current guidelines on *H. pylori* eradication therapy in Korea recommend 14-day therapy as an empirical therapy [4], a real-world study on 14-day tegoprazan-based triple therapy was performed [21]. In this study, the eradication rate was 76.7% in the 14-day tegoprazan group and 75.4% in the 14-day rabeprazole group, without significant differences between the groups. The efficacy of tegoprazan may not have been superior to that of PPI-based therapy due to an insufficient dosage of tegoprazan, which is approved as 100 mg/day (50 mg twice daily) by the Korean Food and Drug Administration. If the dosage of tegoprazan is increased, its eradication rate may be improved since tegoprazan inhibits proton pumps in a dose-dependent manner [22]. When the amount of tegoprazan administered per day was raised from 100 to 200 mg, the median intragastric pH increased from 5.2 to 6.4 [22]. To effectively treat *H. pylori* infection, it may be more beneficial to administer tegoprazan at a dosage of 100 mg twice a day. However, further studies are required to test this hypothesis.

15.3 Conclusion

In this section, we summarize the roles and efficacies of PPIs and P-CABs in *H. pylori* eradication (Table 15.1). PPIs and P-CABs help increase *H. pylori* eradication rates. In particular, vonoprazan has superior efficacy than PPIs in terms of *H. pylori* eradication; however, the efficacy of other P-CABs, including tegoprazan, should be evaluated in future studies.

Table 15.1 Efficacy of P-CAB-based *H. pylori* eradication therapy

Target drug	Country/ region	Study design	Regimen	Control	Main outcome
Vonoprazan	Japan [18]	RCT	7-Day vonoprazan-based triple therapy (vonoprazan 20 mg + amoxicillin 750 mg + clarithromycin 200 or 400 mg, twice daily)	7-Day lansoprazole-based triple therapy (lansoprazole 30 mg + amoxicillin 750 mg + clarithromycin 200 or 400 mg, twice daily)	Vonoprazan triple 92.6%, lansoprazole triple 75.9%, $P < 0.001$
Vonoprazan	The US and Europe [20]	RCT	14-Day vonoprazan-based dual therapy (vonoprazan 20 mg twice daily + amoxicillin 1000 mg thrice daily) 14-Day vonoprazan-based triple therapy (vonoprazan 20 mg + amoxicillin 1000 mg + clarithromycin 500 mg, twice daily)	14-Day lansoprazole-based triple therapy (lansoprazole 30 mg + amoxicillin 1000 mg + clarithromycin 500 mg, twice daily)	Vonoprazan dual 77.2% ($P = 0.013$, vs. lansoprazole triple), vonoprazan triple 80.8% ($P < 0.001$, vs. lansoprazole triple), lansoprazole triple 68.5%
Tegoprazan	Korea [23]	RCT	7-Day tegoprazan-based triple therapy (tegoprazan 50 mg + amoxicillin 1000 mg + clarithromycin 500 mg, twice daily)	7-Day lansoprazole-based triple therapy (lansoprazole 30 mg + amoxicillin 1000 mg + clarithromycin 500 mg, twice daily)	Tegoprazan triple 62.9% (non-inferior to control[a]), lansoprazole triple 60.6%

P-CAB potassium-competitive acid blocker, *RCT* randomized controlled trial
[a]Only *P*-value for the non-inferiority test was reported in this study

References

1. Wagner S, Varrentrapp M, Haruma K, et al. The role of omeprazole (40 mg) in the treatment of gastric Helicobacter pylori infection. Z Gastroenterol. 1991;29:595–8.
2. Lee DH, Park HJ, Song SY, et al. Evaluation of therapeutic regimens for the treatment of Helicobacter pylori infection. Yonsei Med J. 1996;37:270–7.
3. Malfertheiner P, Megraud F, Rokkas T, et al. Management of Helicobacter pylori infection: the Maastricht VI/Florence consensus report. Gut. 2022.
4. Jung HK, Kang SJ, Lee YC, et al. Evidence-based guidelines for the treatment of Helicobacter pylori infection in Korea 2020. Gut Liver. 2021;15:168–95.
5. DeVault KR, Talley NJ. Insights into the future of gastric acid suppression. Nat Rev Gastroenterol Hepatol. 2009;6:524–32.
6. Labenz J, Stolte M, Blum AL, et al. Intragastric acidity as a predictor of the success of Helicobacter pylori eradication: a study in peptic ulcer patients with omeprazole and amoxicillin. Gut. 1995;37:39–43.
7. Sugimoto M, Furuta T, Shirai N, et al. Evidence that the degree and duration of acid suppression are related to Helicobacter pylori eradication by triple therapy. Helicobacter. 2007;12:317–23.
8. Olbe L, Carlsson E, Lindberg P. A proton-pump inhibitor expedition: the case histories of omeprazole and esomeprazole. Nat Rev Drug Discov. 2003;2:132–9.
9. Gedda K, Scott D, Besancon M, Lorentzon P, Sachs G. Turnover of the gastric H+,K(+)-adenosine triphosphatase alpha subunit and its effect on inhibition of rat gastric acid secretion. Gastroenterology. 1995;109:1134–41.
10. Scott DR, Munson KB, Marcus EA, Lambrecht NW, Sachs G. The binding selectivity of vonoprazan (TAK-438) to the gastric H+, K+ -ATPase. Aliment Pharmacol Ther. 2015;42:1315–26.
11. Akazawa Y, Fukuda D, Fukuda Y. Vonoprazan-based therapy for Helicobacter pylori eradication: experience and clinical evidence. Therap Adv Gastroenterol. 2016;9:845–52.
12. Mendlein J, Sachs G. Interaction of a K(+)-competitive inhibitor, a substituted imidazo[1,2a] pyridine, with the phospho- and dephosphoenzyme forms of H+, K(+)-ATPase. J Biol Chem. 1990;265:5030–6.
13. Kirchhoff P, Andersson K, Socrates T, Sidani S, Kosiek O, Geibel JP. Characteristics of the K+-competitive H+,K+-ATPase inhibitor AZD0865 in isolated rat gastric glands. Am J Physiol Gastrointest Liver Physiol. 2006;291:G838–43.
14. Simon WA, Herrmann M, Klein T, et al. Soraprazan: setting new standards in inhibition of gastric acid secretion. J Pharmacol Exp Ther. 2007;321:866–74.

15. Berg AL, Bottcher G, Andersson K, et al. Early stellate cell activation and veno-occlusive-disease (VOD)-like hepatotoxicity in dogs treated with AR-H047108, an imidazopyridine proton pump inhibitor. Toxicol Pathol. 2008;36:727–37.
16. Takahashi N, Take Y. Tegoprazan, a novel potassium-competitive acid blocker to control gastric acid secretion and motility. J Pharmacol Exp Ther. 2018;364:275–86.
17. Suzuki S, Gotoda T, Kusano C, Iwatsuka K, Moriyama M. The efficacy and tolerability of a triple therapy containing a potassium-competitive acid blocker compared with a 7-day PPI-based low-dose clarithromycin triple therapy. Am J Gastroenterol. 2016;111:949–56.
18. Murakami K, Sakurai Y, Shiino M, Funao N, Nishimura A, Asaka M. Vonoprazan, a novel potassium-competitive acid blocker, as a component of first-line and second-line triple therapy for Helicobacter pylori eradication: a phase III, randomised, double-blind study. Gut. 2016;65:1439–46.
19. Jung YS, Kim EH, Park CH. Systematic review with meta-analysis: the efficacy of vonoprazan-based triple therapy on Helicobacter pylori eradication. Aliment Pharmacol Ther. 2017;46:106–14.
20. Chey WD, Megraud F, Laine L, Lopez LJ, Hunt BJ, Howden CW. Vonoprazan triple and dual therapy for Helicobacter pylori infection in the United States and Europe: randomized clinical trial. Gastroenterology. 2022;163:608–19.
21. Jung YS, Kim S, Kim HY, et al. Efficacy and tolerability of 14-day tegoprazan- versus rabeprazole-based triple therapy for eradication of Helicobacter pylori: a real-world evidence study. Gut Liver. 2022.
22. Han S, Choi HY, Kim YH, et al. Randomised clinical trial: safety, tolerability, pharmacokinetics, and pharmacodynamics of single and multiple oral doses of tegoprazan (CJ-12420), a novel potassium-competitive acid blocker, in healthy male subjects. Aliment Pharmacol Ther. 2019;50:751–9.
23. Choi YJ, Lee YC, Kim JM, et al. Triple therapy-based on tegoprazan, a new potassium-competitive acid blocker, for first-line treatment of Helicobacter pylori infection: a randomized, double-blind, phase III, clinical trial. Gut Liver. 2022;16:535–46.

Role of Bismuth

16

Woo Chul Chung

Highlights

What Is Known
• 80~100% of the effects of clarithromycin resistance have resulted in eradication failure. • Adding a bismuth additive to the existing standard triple therapy not only helps overcome the problem of antibiotic resistance but also more than doubles the effectiveness of antibacterial treatment.
Future Research Questions
• It should be confirmed whether bismuth-add-on standard triple therapy serves as an empirical treatment regimen as effective as existing concomitant therapy or bismuth quadruple therapy.

16.1 Introduction

Recently, various results have been published on the change in success rate of *Helicobacter pylori* (*H. pylori*) eradication as the first-line standard triple therapy (STT), but many studies on the trend of eradication rate reduction due to antibiotic resistance have been reported [1–3]. In addition, after the failure of first STT, the eradication rate for classic bismuth quadruple therapy (BQT) as a secondary eradication regimen has also dramatically decreased from more than 90% until 2008 to 70% of 2012 [4, 5]. The reasons for this decrease in eradication rate include antibiotic resistance, age of patients, smoking status, differences in host immunity, underlying diseases, and reduced compliance, but among them, antibiotic resistance is known to be the most important factor. In particular, considering the high resistance rate of clarithromycin, it can be considered as more important factor in result of eradication rate. So far, it has been reported that 80~100% of the effects of clarithromycin resistance have resulted in eradication failure. A high level of resistance to clarithromycin cannot be resolved by increasing the amount of clarithromycin administration or extending the treatment period. Therefore, it is important to find new therapies without the problem of drug resistance.

16.2 Main Body

Bismuth is a chemical element with symbol Bi and atomic number 83. It is classified as a heavy metal, but is known to be of little toxicity. Since ancient times, the compound has been used as a cosmetic or medicine, and is also used for eradication of *H. pylori*. Surprisingly, no stains of *H. pylori* resistant to bismuth have been observed. The proposed mechanism involved in eradication of *H. pylori* is that the bismuth agent forms a complex in the bacterial wall or periplasmic

W. C. Chung (✉)
Department of Internal Medicine, St. Vincent's Hospital, The Catholic University of Korea, Seoul, Korea
e-mail: jwchulkr@catholic.ac.kr

B.-W. Kim (ed.), *Helicobacter Infection in Clinical Practice*,
https://doi.org/10.1007/978-981-97-8397-7_16

space, inhibits the action of enzymes or ATP production, and inhibits the attachment of bacteria to the gastric wall, thereby showing a direct sterilization effect [6]. Furthermore, it is expected that Bismuth will not only promote the improvement of ulcers by enhancing mucosal protective factors such as prostaglandin and epidermal growth factor, but also improve the results of eradication treatment even in strains with confirmed resistance, which are known to cause synergic effects with antibiotics such as metronidazole and clarithromycin [7, 8]. However, when used alone, the eradication effect is low, so it has been used as a combination therapy with other antibiotics, and has been known as a drug mainly used for secondary eradication treatment rather than primary treatment.

According to a recent meta-analysis of the clinical results of adding bismuth to eradication therapy, the eradication success rate was significantly higher when bismuth was added than when bismuth was not added. In particular, it showed superior treatment effects when targeting patients with confirmed resistance or areas with a high rate of clarithromycin-resistant strains. What is noteworthy is that when bismuth was added to the first-line eradication treatment, the treatment effect increased 3.5 times compared to the existing treatment (Table 16.1) [9]. This trend was also found in regions where the frequency of clarithromycin resistance was known to be high, so there was a high need for a prospective study on this, but existing studies lacked consideration for this [10].

In a 2021 domestic study, a total of 107 *H. pylori* first-line eradication patients were treated with Bismuth (tripotassium dicitrate bismuthate, DENOL Greencross Co., Seoul, Korea. 300 mg b.i.d) as a standard three-drug regimen [11]. At the same time, the patients were checked for clarithromycin resistance using polymerase chain reaction (PCR), and the treatment effect of standard triple therapy with bismuth was evaluated in detail according to clarithromycin resistance. Of these, 104 patients could confirm the success of eradication treatment, and it is particularly noteworthy that the success rate of eradication of *H. pylori* was 87.9% in intention-to-treat (ITT) analysis and 90.4% in per-protocol (PP) analysis, respectively. Only for additional information, PCR was conducted to confirm clarithromycin resistance, and the frequency of 23S r RNA point mutation reached 33.6%, and A2143G appeared as the most frequent point mutation. Overall, when analyzed in detail according to clarithromycin resistance, patients without resistance showed a high eradication success rate of 97.1%, and patients with resistance showed a success rate of 77.1%. In this study, it was effective to add bismuth agent to the standard triple therapy treatment to overcome the insufficient clinical effect of the conventionally widely used first-line eradication treatment, and it was clinically acceptable in terms of overcoming clarithromycin resistance. In addition, the frequency of side effects of concern is not so high, so it is expected that it will be an alternative method in actual clinical settings where customized treatment based on clarithromycin resistance test is difficult.

Table 16.1 Clinical trials of bismuth-containing first-line regimens for *Helicobacter pylori* since 2000 (all enrolled studies were limited to randomized control trials)

First author	Year	Nations	Bismuth-based regimens	Control regimens	Diagnosis	Eradication confirm	Duration	Eradication rate	Adverse event rate
G.C.Spinzi	2000	Italy	RBC 400 mg bid AMO 1000 mg bid CLA 500 mg bid	OME 20 mg bid AMO 1000 mg bid CLA 500 mg bid	RUT Histology	RUT Histology After 1 month	7 days	B:60.8%(59/97) C:69.5%(66/95)	B:4.1%(4/97) C:4.2%(4/95)
V.Savarino	2000	Italy	RBC 400 mg bid CLA 250 mg bid MDZ 500 mg bid	OME 20 mg q.d CLA 250 mg bid MDZ 500 mg bid	RUT Histology	RUT Histology After 1 month	7 days	B:84.3%(59/70) C:68.6%(48/70)	B:17.1%(12/70) C:15.7%(11/70)
J.P.Gisbert	2000	Spain	RBC 400 mg bid AMO 1000 mg bid CLA 500 mg bid	PAN 40 mg bid AMO 1000 mg bid CLA 500 g bid	RUT Histology UBT	UBT After 1 month	7 days	B:81.3%(61/75) C:64.0%(48/75)	B:14.7%(11/75) C:17.3%(13/75)
F.Catalano	2000	Italy	RBC 400 mg bid AMO 1000 mg bid CLA 500 mg bid	OME 40 mg qd AMO 1000 mg bid CLA 500 mg bid	RUT Histology	RUT Histology After 1 month	10 days	B:77.8%(42/54) C:81.8%(45/55)	B:42.6%(23/54) C:45.5%(25/55)
H.V.Mao	2000	Vietnam	RBC 400 mg bid AMO 1000 mg bid CLA 500 mg bid	OME 20 mg bid AMO 1000 mg bid CLA 500 mg bid	RUT Histology	RUT Histology After 1 month	10 days	B:91.4%(43/47) C:86.0%(49/57)	B:14.9%(7/47) C:16.9%(10/59)
F.Meguraud	2000	UK	RBC 400 mg bid CLA 500 mg bid	OME 20 mg bid CLA 500 mg bid	RUT Histology Culture UBT	Histology UBT Culture After 1 month	14 days	B:88.7%(63/71) C:52.7%(39/74)	NA
L.Bujanda	2001	Spain	RBC 400 mg bid AMO 1000 mg bid CLA 500 mg bid	OME 20 mg bid AMO 1000 mg bid CLA 500 mg bid	RUT UBT	UBT After 1 month	7 days	B:82.7%(43/52) C:72.3%(34/47)	B:42.3%(22/52) C:44.7%(21/47)
B.C.Y.Wong	2001	Hong Kong	RBC 400 mg bid CLA 250 mg bid MDZ 400 mg bid	OME 20 mg bid CLA 250 mg bid MDZ 400 mg bid	RUT Histology Culture	RUT CLO Histology Culture After 6 weeks	7 days	B:83.3%(75/90) C:65.5%(59/90)	B:76.7%(69/90) C:55.5%(50/90)
C.H.Chuang	2001	Taiwan	RBC 400 mg bid AMO 1000 mg bid MDZ 400 mg bid	OME 20 mg bid AMO 1000 mg bid MDZ 400 mg bid	RUT Histology UBT	UBT After 6 weeks	7 days	B: 70.0%(21/30) C:68.9%(20/29)	NA

(continued)

Table 16.1 (continued)

First author	Year	Nations	Bismuth-based regimens	Control regimens	Diagnosis	Eradication confirm	Duration	Eradication rate	Adverse event rate
W.K.Hung	2002	China	RBC 400 mg bid AMO 1000 mg bid MDZ 400 mg bid	OME 20 mg bid AMO 1000 mg bid MDZ 400 mg bid	RUT Histology	UBT After 5 weeks	7 days	B:77.3%(92/119) C:77.3%(85/110)	B:10.0%(12/119) C:25.4%(28/110)
P.G.Farup	2002	Norway	RBC 400 mg bid CLA 250 mg bid MDZ 500 mg bid	OME 20 mg bid CLA 250 mg bid MDZ 500 mg bid	RUT UBT	UBT After 7–14 weeks	7 days	B:96.6%(141/146) C:85.4%(117/137)	B:6.8% (10/146) C:6.6% (9/137)
C.Severi	2009	Italy	TDB 240 mg bid AMO 1000 mg tid MDZ 250 mg tid	LAN 30 mg q.d AMO 1000 mg tid MDZ 250 mg tid	Histology	UBT After 3 months	14 days	B:75.9%(41/54) C:46.3%(25/54)	NA
M.Demir	2009	Turkey	RBC 400 mg bid AMO 1000 mg bid CLA 500 mg bid	PAN 40 mg bid AMO 1000 mg bid CLA 500 mg bid	RUT Histology	UBT After 6 weeks	14 days	B:61.7% (66/107) C:63.6% (68/107)	B: 18.7%(20/107) C:13.1% (14/107)
E.Avsar	2013	Turkey	RBC 400 mg bid AMO 1000 mg bid CLA 500 mg bid	LAN 30 mg bid AMO 1000 mg bid CLA 500 mg bid	RUT Histology	RUT After 6 weeks	14 days	B: 65.1%(97/149) C: 63.8%(83/130)	B:54.4%(81/149) C:56.9%(74/130)
L.Wang	2017	China	BSC 120 mg qid OME 20 mg bid AMO 1000 mg bid CLA 500 mg bid	OME 20 mg bid AMO 1000 mg bid CLA 500 mg bid	Histology UBT	Histology UBT After 3 months	10 days	B:86.1%(143/166) C:58.4%(108/185)	NA
O.Ozturk	2017	Turkey	BSC 600 mg bid OME 20 mg bid AMO 1000 mg bid CLA 500 mg bid	OME 20 mg bid AMO 1000 mg bid CLA 500 mg bid	Histology UBT	UBT After 6–8 weeks	10 days	B:76.5%(39/51) C:61.2%(30/49)	B:43.1%(22/51) C:40.8%(20/49)
X.Long	2018	China	B.P.C 600 mg bid ESO 20 mg bid CLA 500 mg bid MDZ 400 mg qid	ESO 20 mg bid CLA 500 mg bid MDZ 400 mg qid	RUT Histology Culture UBT	UBT After 6 weeks	14 days	B:84.8%(28/33) C:63.6%(21/33)	B:48.5%(16/33) C:45.5%(15/33)

AMO amoxicillin, *MDZ* metronidazole, *CLA* clarithromycin, *BSC* bismuth subcitrate, *BSN* bismuth subnitrate, *TDB* tripostassium dicitrate bismuthate, *RBC* ranitidine bismuth citrate, *BPC* bismuth potassium citrate, *OME* omeprazole, *LAN* lansoprazole, *RAB* rabeprazole, *ESO* esomezole, *PAN* pantoprazole, *RUT* rapid urease test, *UBT* urea breath test, *B* bismuth group, *C* control group, *NA* not applicable

16.3 Conclusion

As the number of strains resistant to antibiotics increases, the treatment success rate is gradually decreasing. Accordingly, a method of changing the antibiotic combination to a new form has been suggested as a recommendation, but this alternative also has the limitation that it can cause new antibiotic resistance. Previous studies have confirmed that the existing standard triple therapy with bismuth add-on can overcome the limitations of recommending a new antibiotic combination in the treatment of *H. pylori* eradication. It is expected that it will serve as an important foundation for future prospective research on *H. pylori* eradication therapy.

References

1. Gong EJ, Yun SC, Jung HY, et al. Meta-analysis of first-line triple therapy for Helicobacter pylori eradication in Korea: is it time to change? J Korean Med Sci. 2014;29:704–13.
2. Shin WG, Lee SW, Baik GH, et al. Eradication rates of Helicobacter pylori in Korea over the past 10 years and correlation of the amount of antibiotics use: nationwide survey. Helicobacter. 2016;21:266–78.
3. Gong EJ, Ahn JY. Antimicrobial resistance of Helicobacter pylori isolates in Korea. Kor J Helicobacter Upper Gastrointest Res. 2018;18:82–8.
4. Kim J, Kim YJ, Chung WC. Proton pump inhibitor switching strategy after failure of standard triple therapy for Helicobacter pylori Infection. Kor J Helicobacter Upper Gastrointest Res. 2020;20:146–52.
5. Kim GH, Kim JA, Ko UW, et al. Eradication rates of first-line and second-line therapy for Helicobacter pylori infection in Gyeongnam province. Kor J Helicobacter Upper Gastrointest Res. 2015;15:160–5.
6. Dore MP, Graham DY, Mele R, et al. Colloidal bismuth subcitrate-based twice-a-day quadruple therapy as primary or salvage therapy for Helicobacter pylori infection. Am J Gastroenterol. 2002;97:857–60.
7. Gomollón F, Ducóns JA, Ferrero M, et al. Quadruple therapy is effective for eradicating Helicobacter pylori after failure of triple proton-pump inhibitor-based therapy: a detailed, prospective analysis of 21 consecutive cases. Helicobacter. 1999;4:222–5.
8. Wu DC, Hsu PI, Chen A, et al. Randomized comparison of two rescue therapies for Helicobacter pylori infection. Eur J Clin Invest. 2006;36:803–9.
9. Ko SW, Kim YJ, Chung WC, Lee SJ. Bismuth supplements as the first-line regimen for Helicobacter pylori eradication therapy: systemic review and meta-analysis. Helicobacter. 2019;24:e12565.
10. McNicholl AG, Bordin DS, Lucend A, et al. Combination of bismuth and standard triple therapy eradicates Helicobacter pylori infection in more than 90% of patients. Clin Gastroenterol Hepatol. 2020;18:89–98.
11. Kim YJ, Chung WC, Kim DB. Efficacy of bismuth added to standard triple therapy as the first-line eradication regimen for Helicobacter pylori infection. Helicobacter. 2021;26:e12792.

Part IV

Guidelines for *Helicobacter pylori*

17 Comparison of Recent *Helicobacter pylori* Guidelines

Joon Sung Kim and Jin Il Kim

Highlights

What Is Known
• Antibiotic susceptibility testing (molecular or culture) should be considered before prescribing treatment. • Bismuth quadruple therapy should be considered as empirical treatment in areas of high clarithromycin resistance. • Clarithromycin-based triple therapy can be considered an empirical treatment in areas of low clarithromycin resistance. • Regimens containing unneeded antibiotics should be avoided.
Future Research Questions
• Optimized guidelines should be developed while considering regional-specific requirements.

17.1 Introduction

Helicobacter pylori (*H. pylori*) can cause peptic ulcer disease and gastric cancer [1]. *H. pylori* is present in more than half of the world's population, and the overall seroprevalence of *H. pylori* infection is 51.0% in South Korea [2, 3]. Recently, a paradigm shift regarding *H. pylori* has occurred. First, the Kyoto consensus report designated *H. pylori* gastritis as an infectious disease [4]. *H. pylori* gastritis is now a nosological entity in the new International Classification of Disease 11th Revision (ICD 11). The recent Maastricht VI consensus stated *H. pylori* infection always causes gastritis, irrespective of symptoms or complications [5]. It also stated that *H. pylori* is a gastric pathogen and *H. pylori* gastritis is an infectious disease. The indication for treatment is no longer reserved for patients with clinical manifestations of infection; treatment is required in all *H. pylori*-infected patients.

Gastric cancer is the fourth most common cause of cancer-related mortality worldwide [6]. Almost half of the incident cases and deaths occur in East Asia. In 2019, the age-standardized incidence rate of gastric cancer per 100,000 was 28.67 in South Korea [7]. The recent Taipei consensus recommends screening and treating *H. pylori* in countries with a high incidence of gastric cancer [8]. Combined, these statements drive arguments that *H. pylori* should be treated

J. S. Kim
Department of Internal Medicine, Incheon St. Mary's Hospital, The Catholic University of Korea, Seoul, Korea
e-mail: kijoons@catholic.ac.kr

J. I. Kim (✉)
Department of Internal Medicine, Yeouido St. Mary's Hospital, The Catholic University of Korea, Seoul, Korea
e-mail: jikim@catholic.ac.kr

B.-W. Kim (ed.), *Helicobacter Infection in Clinical Practice*,
https://doi.org/10.1007/978-981-97-8397-7_17

in all infected patients. However, controversy persists on whether the general population should be screened and treated for *H. pylori* in Korea. This is due to the high prevalence of infection, limited resources, and unknown cost-effectiveness for eradicating every patient. In 2018, the South Korean government insurance system started to cover eradication therapy for *H. pylori* gastritis. In 2020, the Korean College of Helicobacter and Upper Gastrointestinal Research group published the third edition of the Korean guidelines on treating *H. pylori* infection [9].

The prevalence of *H. pylori* infection and gastric cancer is relatively low in the United States and Europe [10]. However, managing *H. pylori* infection is still an area of interest in these areas as gastric cancer is frequently diagnosed at an advanced state, and mortality rates from gastric cancer are still high. The recently revised clinical guidelines for *H. pylori* management from *the* American College of Gastroenterology (ACG) were reported in 2017 [11]. The most significant development in this guideline was regarded in the medical treatment of *H. pylori* infection, and clinicians have been able to test various regimens for eradication. The American Gastroenterological Association (AGA) clinical practice update was published in 2021 [12]. The AGA update review only covered the management of refractory *H. pylori* infection.

The Maastricht VI/Florence Consensus report was published in 2022 [5]. The latest Maastricht VI guidelines recommend routinely performing susceptibility tests if available, even before prescribing first-line treatment concerning antibiotic stewardship. However, it remains to be established if this strategy is practical in clinical practice. Whether all subjects receiving an upper endoscopy must provide samples to determine antibiotic susceptibility is debatable. In this chapter, we aimed to investigate the current recommendation concerning the indications and treatment of *H. pylori* based on the abovementioned guidelines.

17.2 Main Body

17.2.1 Indications

The recent Korean guidelines recommend five strong and four weak recommendations for eradication [9]. The five-strong recommendations are peptic ulcer diseases, gastric MALT lymphoma, after endoscopic resection of early gastric cancer (EGC), long-term low-dose aspirin user with a history of peptic ulcer diseases, and idiopathic thrombocytopenic purpura (ITP). The weak recommendations are family history of gastric cancer, functional dyspepsia, unexplained iron-deficiency anemia (IDA), and after endoscopic resection of gastric adenoma. The recent Korean guidelines dismissed atrophic gastritis and intestinal metaplasia due to a lack of evidence and expert agreement.

The insurance coverage policy is another issue in Korea. Until 2017, the Korean Health Insurance Review and Assessment Service allowed eradication only in patients with peptic ulcer diseases, low-grade gastric MALT lymphoma, and after endoscopic resection of EGC. However, the indication for eradication was expanded on January 1, 2018. Patients diagnosed with ITP after endoscopic resection of gastric adenoma, first-degree relatives of gastric cancer, atrophic gastritis when medical judgment requires eradication therapy, and when the patient agreed to administer the treatment were included. Practically, all patients diagnosed with *H. pylori* gastritis can be treated based on the changes in Korean Health insurance. Thus, the proportion of eradication treatment is predicted to increase in the near future, but it remains to be seen if this leads to decreased incidence of gastric cancer in Korea.

The ACG clinical guidelines divided indications into two broad categories—indications to test and indications to treat [11]. The ACG guidelines strongly recommend treating all patients with active infection, and those with the crucial problem should be tested for infection. The four established and recommended indications for diagnosis and eradication include active peptic ulcer diseases, history of peptic ulcer diseases, low-grade gastric MALT lymphoma, and after endoscopic resection of EGC. Patients receiving treatment with chronic nonsteroidal anti-inflammatory (NSAID) drug was also a strong recommendation for testing and treatment. Patients with dyspeptic symptoms and typical gastroesophageal reflux disease (GERD) symptoms are also strongly recommended for eradication. However, the indications to test for *H. pylori* in these patients differed slightly from those to treat. The conditional indications are long-term, low-dose aspirin use, unexplained IDA, and ITP. Noteworthy, there are no recommendations for routine diagnosis and treatment in patients with lymphocytic gastritis, hyperplastic gastric polyps, and asymptomatic patients with a family history of gastric cancer.

The Maastricht VI guidelines include 16 statements regarding indications and associations of *H. pylori* [5]. Indications for eradication include dyspepsia, high-risk patients on long-term aspirin, and naïve patients starting long-term NSAID therapy. They also recommend eradication in patients with unexplained IDA, ITP, vitamin B12 deficiency, and low-grade gastric MALT lymphoma.

17.2.2 Initial Treatment

The Korean guidelines recommend primary treatment with 2 weeks of standard clarithromycin triple therapy comprising a standard dose of proton pump inhibitor (PPI) (1 g), amoxicillin (1 g), and clarithromycin (500 mg) twice a day [9]. A 7-day clarithromycin triple therapy as first-line treatment is only recommended after performing a PCR or sequencing test to confirm clarithromycin resistance. Sequential therapy and concomitant therapy of 10 days are also recommended as first-line treatments in the new guidelines. Sequential therapy consists of a 5-day standard dose of PPI and 1 g amoxicillin twice daily, followed by another 5-day standard dose of PPI, 500 mg clarithromycin, and 500 mg metronidazole twice daily. Concomitant therapy consists of 500 mg metronidazole twice daily, included in standard triple therapy. Bismuth quadruple therapy is also recommended as a first-line eradication therapy; however, it should only be used when other therapies are unfeasible because of its high adverse effects and the possibility of its use as salvage therapy.

The ACG guidelines recommend more various regimens than the Korean guidelines for first-line eradication therapy [11]. Fourteen-day clarithromycin triple therapy is recommended in areas with low rates of clarithromycin resistance (<15%) and patients without a previous history of macrolide exposure. Additionally, 10- to 14-day bismuth quadruple therapy, 10- to 14-day sequential therapy, 10- to 14-day concomitant therapy, 14-day hybrid therapy, 10- to 14-day levofloxacin triple therapy, and 10- to 14-day fluoroquinolone sequential therapy are all recommend as first-line regimens. The regimens of clarithromycin triple, bismuth quadruple, sequential, concomitant, and levofloxacin triple therapies are identical to those in the Korean guidelines. Hybrid therapy consists of a standard PPI dose and 1 g amoxicillin for 7 days twice a day, followed by an additional 7-day standard PPI dose, 1 g amoxicillin, 500 mg clarithromycin, and 500 mg nitroimidazole twice a day. Levofloxacin sequential therapy was similar to the sequential therapy, with the drug changed from 500 mg clarithromycin twice a day to 500 mg levofloxacin once a day, and the duration extended from 10 days to 10–14 days.

The Maastricht guidelines recommend susceptibility tests (molecular or culture) before prescribing first-line treatment [5]. If susceptibility testing is unavailable, clarithromycin triple therapy or bismuth quadruple therapy of 14 days is recommended in areas with low clarithromycin resistance. In areas of high resistance, bismuth quadruple therapy is recommended, and concomitant therapy may be considered if bismuth quadruple therapy is not available. These are summarized in Table 17.1.

Table 17.1 First-line treatment recommendations for *H. pylori* eradication in the different guidelines

Korea [9]	ACG [11]	Maastricht [5]
Clarithromycin triple Sequential Concomitant Bismuth quadruple	Clarithromycin triple Sequential Concomitant Bismuth quadruple Hybrid Levofloxacin triple Levofloxacin sequential LOAD[a]	Low resistance (<15%) Clarithromycin triple Bismuth quadruple High resistance (>15%) Bismuth quadruple Concomitant

[a]Proton-pump inhibitor, levofloxacin, nitazoxanide, doxycycline

17.2.3 Second-Line Therapy

The Korean guidelines recommend 2-week bismuth quadruple therapy as a second-line regimen in patients with failure of clarithromycin triple therapy or non-bismuth quadruple therapy (sequential or concomitant therapy) [9]. For second-line therapy, the ACG guidelines proposed the following five regimens for use: 14-day bismuth quadruple therapy, 14-day levofloxacin triple therapy, 10- to 14-day concomitant therapy, 10-day rifabutin triple therapy, and 14-day high-dose dual therapy [11]. After the failure of first-line clarithromycin-containing therapy, bismuth quadruple or levofloxacin-containing regimens were recommended. After first-line bismuth quadruple therapy failed, clarithromycin-containing or levofloxacin-containing regimens were recommended. However, the ACG guidelines recommended that clarithromycin triple therapy should not be prescribed as salvage therapy.

The AGA clinical practice update recommends levofloxacin triple therapy, rifabutin triple therapy, and an alternative bismuth-containing quadruple therapy, such as PBLA (PPI, bismuth, levofloxacin, amoxicillin), PBLT (PPI, bismuth, levofloxacin, tetracycline), PBLM (PPI, bismuth, levofloxacin, metronidazole), or PBCT (PPI, bismuth, clarithromycin, tetracycline) after failure of bismuth quadruple therapy [12].

The Maastricht guidelines recommend that local resistance patterns guide empiric second-line and rescue therapy [5]. After the failure of a clarithromycin triple therapy or bismuth quadruple therapy, a fluoroquinolone-containing quadruple (or triple) therapy or a high-dose dual therapy consisting of PPI and amoxicillin is recommended as a second-line treatment. Combining bismuth with other antibiotics or rifabutin may be an option in areas of high fluoroquinolone resistance. After a non-bismuth quadruple therapy fails, either a bismuth quadruple therapy or a fluoroquinolone-containing quadruple (or triple) therapy is recommended. PPI-amoxicillin high-dose dual therapy might also be considered.

17.2.4 Third-Line Therapy

The Korean guidelines recommend levofloxacin triple therapy, which consists of a standard dose of PPI twice daily, 500 mg levofloxacin once daily, or 250 mg levofloxacin twice daily, and 1 g amoxicillin twice daily as third-line therapy [9]. The other suggested regimens included triple therapy containing other fluoroquinolones, fluoroquinolone quadruple therapy, rifabutin-containing therapy, high-dose dual therapy, or concomitant therapy.

The ACG guideline recommends high-dose dual therapy, levofloxacin quadruple therapy, such as PBLT (PPI, bismuth, levofloxacin, tetracycline) or PBLM (PPI, bismuth, levofloxacin, metronidazole), and repeat PBMT (PPI, bismuth, metronidazole, tetracycline) with high-dose metronidazole (1.5–2 g) divided a day for third-line therapy [11].

The Maastricht guidelines recommend fluoroquinolone regimens after the failure of first-line treatment with clarithromycin triple or non-bismuth quadruple therapies and second-line with bismuth quadruple therapy [5]. In regions with a known high fluoroquinolone resistance, a bismuth quadruple therapy with different antibiotics, rifabutin-containing rescue therapy, or a high-dose PPI-amoxicillin dual therapy, is recommended as a third-line treatment. After the failure of the first-line treatment with clarithromycin or non-bismuth quadruple therapies, and second-line treatment with fluoroquinolone-containing therapy, bismuth quadruple therapy is recommended as third-line treatment. If bismuth is unavailable, high-dose PPI-amoxicillin dual or a rifabutin-containing regimen could be considered. In areas with low clarithromycin resistance, a clarithromycin-based triple or quadruple therapy/high dose dual therapy, a rifabutin-containing regimen, or a combination of bismuth with different antibiotics is recommended.

17.3 Conclusion

The recent availability of susceptibility testing for *H. pylori* has resulted in paradigm shifts in the diagnosis and treatment. This is apparent in the recent Korean and Maastricht guidelines, which recommend susceptibility testing as a treatment option. Regimens containing clarithromycin and metronidazole are still recommended as empirical treatment in Korea. However, recent reviews recommend that clarithromycin, metronidazole, and levofloxacin-based regimens should only administered as susceptibility-based therapy [13]. Including sequential and concomitant therapy in the Korean guidelines may be considered obsolete based on antibiotic stewardship, suggesting that regimens containing unneeded antibiotics should not be given.

The recent paradigm shift recommends that *H. pylori* infections be treated similarly to other infectious diseases. Global resistance to frequently used antibiotics such as clarithromycin, metronidazole, and fluoroquinolone with the availability of susceptibility testing has resulted in a paradigm shift for treatment. Optimizing the current treatment combinations by achieving optimal acid suppression using potassium channel acid blockers needs exploration. In future clinical practice, these should be considered to implement the most effective and practical approach to treating *H. pylori*.

References

1. Crowe SE. Helicobacter pylori infection. N Engl J Med. 2019;380:1158–65.
2. Hooi JKY, Lai WY, Ng WK, et al. Global prevalence of Helicobacter pylori infection: systematic review and meta-analysis. Gastroenterology. 2017;153:420–9.
3. Lee JH, Choi KD, Jung HY, et al. Seroprevalence of Helicobacter pylori in Korea: a multicenter, nationwide study conducted in 2015 and 2016. Helicobacter. 2018;23:e12463.
4. Sugano K, Tack J, Kuipers EJ, et al. Kyoto global consensus report on Helicobacter pylori gastritis. Gut. 2015;64:1353–67.
5. Malfertheiner P, Megraud F, Rokkas T, et al. Management of Helicobacter pylori infection: the Maastricht VI/Florence consensus report. Gut. 2022.
6. Sung H, Ferlay J, Siegel RL, et al. Global Cancer Statistics 2020: GLOBOCAN estimates of incidence and mortality worldwide for 36 cancers in 185 countries. CA Cancer J Clin. 2021;71:209–49.
7. Yang X, Zhang T, Zhang H, Sang S, Chen H, Zuo X. Temporal trend of gastric cancer burden along with its risk factors in China from 1990 to 2019, and projections until 2030: comparison with Japan, South Korea, and Mongolia. Biomark Res. 2021;9:84.
8. Liou JM, Malfertheiner P, Lee YC, et al. Screening and eradication of Helicobacter pylori for gastric cancer prevention: the Taipei global consensus. Gut. 2020;69:2093–112.
9. Jung HK, Kang SJ, Lee YC, et al. Evidence-based guidelines for the treatment of Helicobacter pylori infection in Korea 2020. Gut Liver. 2021;15:168–95.
10. Pounder RE, Ng D. The prevalence of Helicobacter pylori infection in different countries. Aliment Pharmacol Ther. 1995;9(Suppl 2):33–9.
11. Chey WD, Leontiadis GI, Howden CW, Moss SF. ACG clinical guideline: treatment of Helicobacter pylori infection. Am J Gastroenterol. 2017;112:212–39.
12. Shah SC, Iyer PG, Moss SF. AGA clinical practice update on the management of refractory Helicobacter pylori infection: expert review. Gastroenterology. 2021;160:1831–41.
13. Graham DY. Implications of the paradigm shift in management of Helicobacter pylori infections. Therap Adv Gastroenterol. 2023;16:17562848231160858.

MIX
Papier aus verantwortungsvollen Quellen
Paper from responsible sources
FSC® C105338

If you have any concerns about our products,
you can contact us on
ProductSafety@springernature.com

In case Publisher is established outside the EU,
the EU authorized representative is:
Springer Nature Customer Service Center GmbH
Europaplatz 3, 69115 Heidelberg, Germany

Printed by Libri Plureos GmbH
in Hamburg, Germany